STUDENT SOLUTIONS MANUAL

NANCY S. BOUDREAU
Bowling Green State University

A FIRST COURSE IN Statistics

SIXTH EDITION

JAMES T. MCCLAVE / TERRY SINCICH

PRENTICE HALL, UPPER SADDLE RIVER, NJ 07458

Acquisition Editor: ***Millicent Treloar***
Production Editor: ***Ann Marie Longobardo***
Special Projects Manager: ***Barbara A. Murray***
Supplement Cover Manager: ***Paul Gourhan***
Manufacturing Buyer: ***Alan Fischer***

Simon & Schuster / A Viacom Company
Upper Saddle River, NJ 07458

Printed in the United States of America

10 9 8 7 6 5 4 3

ISBN 0-13-595539-4

Prentice-Hall International (UK) Limited, *London*
Prentice-Hall of Australia Pty. Limited, *Sydney*
Prentice-Hall Canada Inc., *Toronto*
Prentice Hall Hispanoamericana, S.A., *Mexico*
Prentice-Hall of India Private Limited, *New Delhi*
Prentice-Hall of Japan, Inc., *Tokyo*
Simon & Schuster Asia Pte. Ltd., *Singapore*
Editoria Prentice-Hall do Brasil, *Ltda., Rio de Janero*

Contents

Preface

This solutions manual is designed to accompany the text, ***A First Course in Statistics***, Sixth Edition, by James T. McClave, Frank H. Dietrich, II, and Terry Sincich. It provides answers to most odd-numbered exercises for each chapter in the text. Other methods of solution may also be appropriate; however, the author has presented one that she believes to be most instructive to the beginning Statistics student. The student should first attempt to solve the assigned exercises without help from this manual. Then, if unsuccessful, the solution in the manual will clarify points necessary to the solution. The student who successfully solves an exercise should still refer to the manual's solution. Many points are clarified and expanded upon to provide maximum insight into and benefit from each exercise.

Instructors will also benefit from the use of this manual. It will save time in preparing presentations of the solutions and possibly provide another point of view regarding their meaning.

Some of the exercises are subjective in nature and thus omitted from the Answer Key at the end of ***A First Course in Statistics***, Sixth Edition. The subjective decisions regarding these exercises have been made and are explained by the author. Solutions based on these decisions are presented; the solution to this type of exercise is often most instructive. When an alternative interpretation of an exercise may occur, the author has often addressed it and given justification for the approach taken.

I would like to thank Kelly Evans for creating the art work and Brenda Dobson for her assistance and for typing this work.

Nancy S. Boudreau
Bowling Green State University
Bowling Green, Ohio

Statistics, Data, and Statistical Thinking

Chapter 1

1.1 Statistics is a science that deals with the collection, classification, analysis, and interpretation of information or data. It is a meaningful, useful science with a broad, almost limitless scope of applications to business, government, and the physical and social sciences.

1.3 The first element of inferential statistics is the population of interest. The population is a set of existing units. The second element is one or more variables that are to be investigated. A variable is a characteristic or property of an individual population unit. The third element is the sample. A sample is a subset of the units of a population. The fourth element is the inference about the population based on information contained in the sample. A statistical inference is an estimate, prediction, or generalization about a population based on information contained in a sample. The fifth and final element of inferential statistics is the measure of reliability for the inference. The reliability of an inference is how confident one is that the inference is correct.

1.5 Quantitative data are measurements that are recorded on a meaningful numerical scale. Qualitative data are measurements that are not numerical in nature; they can only be classified into one of a group of categories.

1.7 A population is a set of existing units such as people, objects, transactions, or events. A sample is a subset of the units of a population.

1.9 An inference without a measure of reliability is nothing more than a guess. A measure of reliability separates statistical inference from fortune telling or guessing. Reliability gives a measure of how confident one is that the inference is correct.

1.11 The data consisting of the classifications A, B, C, and D are qualitative. These data are nominal and thus are qualitative. After the data are input as 1, 2, 3, and 4, they are still nominal and thus qualitative. The only differences between the two data sets are the names of the categories. The numbers associated with the four groups are meaningless.

1.13 a. The population of interest is all citizens of the United States.

b. The variable of interest is the view of each citizen as to whether the president is doing a good or bad job. It is qualitative.

c. The sample is the 2000 individuals selected for the poll.

d. The inference of interest is to estimate the proportion of all citizens who believe the president is doing a good job.

e. The method of data collection is a survey.

f. It is not very likely that the sample will be representative of the population of all citizens of the United States. By selecting phone numbers at random, the sample will be limited to only those people who have telephones. Also, many people share the same phone number, so each person would not have an equal chance of being contacted. Another possible problem is the time of day the calls are made. If the calls are made in the evening, those people who work in the evening would not be represented.

1.15 a. The bacteria count is a number. Therefore, it is quantitative.

b. Occupations take on values such as doctor, lawyer, carpenter, etc., which are not numeric. Therefore, it is qualitative.

c. Marital status can take on values such as married, single, divorced, etc., which are not numeric. Therefore, it is qualitative.

d. The time in months can take on values such as 1, 2, 3, etc. Therefore, it is quantitative.

1.17 a. Length of maximum span can take on values such as 15 feet, 50 feet, 75 feet, etc. Therefore, it is quantitative.

b. The number of vehicle lanes can take on values such as 2, 4, etc. Therefore, it is quantitative.

c. The answer to this item is "yes" or "no," which are not numeric. Therefore, it is qualitative.

d. Average daily traffic could take on values such as 150 vehicles, 3,579 vehicles, 53,295 vehicles, etc. Therefore, it is quantitative.

e. Condition can take on values "good," "fair," or "poor," which are not numeric. Therefore, it is qualitative.

f. The length of the bypass or detour could take on values such as 1 mile, 4 miles, etc. Therefore, it is quantitative.

g. Route type can take on values "interstate," U.S.," "state," "county," or "city," which are not numeric. Therefore, it is qualitative.

1.19 a. The population from which the sample was selected is the set of all department store executives.

b. There are two variables measured by the authors. They are job-satisfaction and Machiavellian rating for each of the executives.

c. The sample is the set of 218 department store executives who completed the questionnaire.

d. The method of data collection is a survey.

e. The inference made by the authors is that those executives with higher job-satisfaction scores are likely to have a lower 'mach' rating.

1.21 a. The population of interest to the psychologist is the set of all male soccer players between the ages of 14 and 29 who played up to five times per week.

b. The variables of interest are the frequency of "headers" and the IQ of each soccer player.

c. The average number of headers per game per player can take on values such as 0, 1, 2, etc. Therefore, it is quantitative. IQ can take on values such as 85, 100, 103, etc. Therefore, it is quantitative.

d. The sample is the set of 60 male soccer players between the ages of 14 and 29 who played up to five times per week.

e. The inference made by the psychologist is that "heading" the ball in soccer lowers players' IQs.

f. There could be several reasons why the inference could be misleading. First, "heading" the ball is a skill. It is possible that those players who are more athletically inclined are not as intellectually gifted as those players who are not as athletically inclined. Also, it appears that the data were collected only once per player. The only way one could conclude that "heading" the ball lowers players' IQs is by collecting data on each player over a period of time and seeing if each player's IQ decreases.

1.23 a. The population of interest is the set of all possible 2-milliliters of a basic solution.

b. The variable of interest is the amount of hydrochloric acid necessary to neutralize 2-milliliters of a basic solution. Since this amount is numeric, the variable is quantitative.

c. The sample is the set of five 2-milliliter portions of the basic solution prepared by the student.

d. The method of data collection is by designed experiment.

1.25 a. The population of interest to the pollsters is the set of all Americans.

b. The variable of interest is whether or not the person believes the Nazi extermination of the Jews happened. Since the views are not numeric, the variable is qualitative.

c. The sample is the set of 1000 adults and high school students who were asked the question.

d. The method of data collection is by survey.

e. The pollsters inferred that one in five Americans believe that it is possible that the Holocaust never happened.

f. The reliability of this inference is very suspect. The question presented to the sample is not a "yes," "no" question. It asks one to decide if it seems possible or if it seems impossible. Therefore, an answer of "yes" to this question is totally meaningless.

Methods for Describing Sets of Data

Chapter 2

2.1 First, we find the frequency of the grade A. The sum of the frequencies for all 5 grades must be 200. Therefore, subtract the sum of the frequencies of the other 4 grades from 200. The frequency for grade A is:

$$200 - (36 + 90 + 30 + 28) = 200 - 184 = 16$$

To find the relative frequency for each grade, divide the frequency by the total sample size, 200. The relative frequency for the grade B is 36/200 = .18. The rest of the relative frequencies are found in a similar manner and appear in the table:

Grade on Statistics Exam	Frequency	Relative Frequency
A: 90–100	16	.08
B: 80– 89	36	.18
C: 65– 79	90	.45
D: 50– 64	30	.15
F: Below 50	28	.14
Total	200	1.00

2.3 The pie chart for these data is:

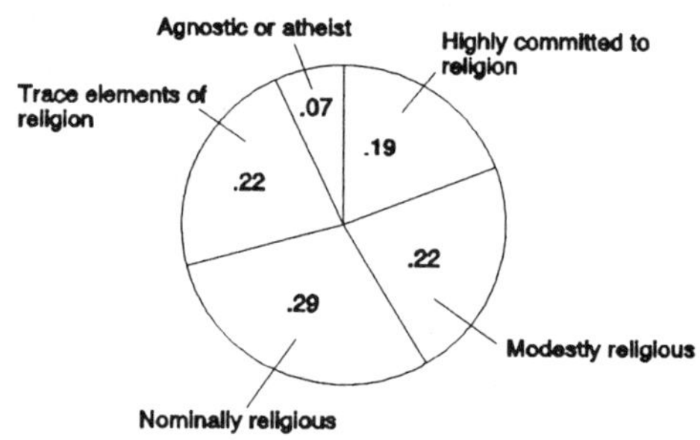

When asked how religious they were, more people responded "Nominally religious" than any other response. About the same number responded to each of the following categories: "Highly committed to religion," "Modestly religious," and "Trace elements of religion." The fewest number responded "Agnostic or atheist."

2.5 A relative frequency bar graph is used to depict the data:

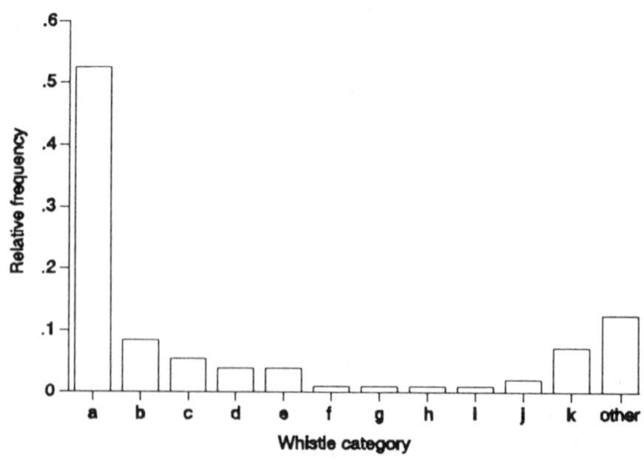

From the bar graph, over half of the whistle types were "Type a." The next most frequent category was the "Other types" with a relative frequency of about .14. Whistle types b and k were the next most frequent. None of the other whistle types had relative frequencies higher than .05.

2.7 a. To compute the relative frequencies for each category, divide the frequency by the total sample size. For category "Violence to others," the relative frequency is 373/562 = .66. The rest of the relative frequencies are found in a similar manner and are displayed in the table.

Media Coverage	Number of Items	Relative Frequency
Violence to others	373	.66
Sympathetic	102	.18
Harm to self	71	.13
Comic images	12	.02
Criticism of definitions	4	.01
Total	562	1.00

b. The relative frequency bar graph is:

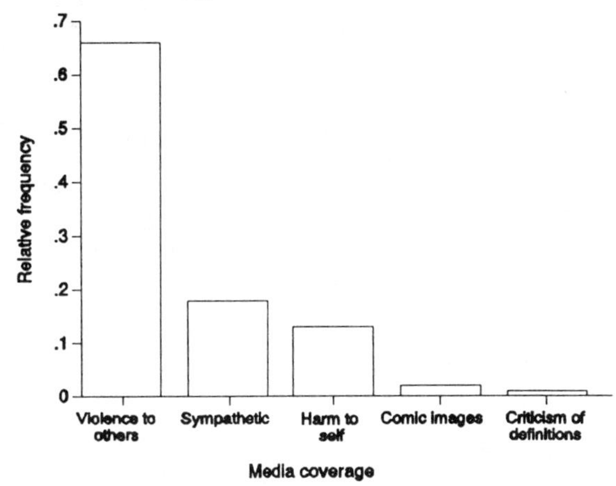

2.9 a. From the bar chart, there are approximately 195 transgenic plant trials approved for herbicide tolerance over this period.

b. From the bar chart, there are approximately 90 transgenic plant trials approved for developing virus resistant crops over this period.

c. To modify the bar graph to show relative frequencies, we must first find the total number of transgenic plant trials over the period. From the bar graph, the following are the approximate frequencies:

Trait	Frequency	Relative Frequency
Unspecified	9	9/550 = .016
Multiple Traits	13	13/550 = .024
Herbicide Tolerance	195	195/550 = .355
Crop Production	180	180/550 = .327
Product Quality	103	103/550 = .187
Marker Genes	50	50/550 = .091
Totals	550	1.000

To find the relative frequencies, divide the frequency by the total sample size, 550. The relative frequencies appear in the table above.

The bar chart showing the relative frequencies is:

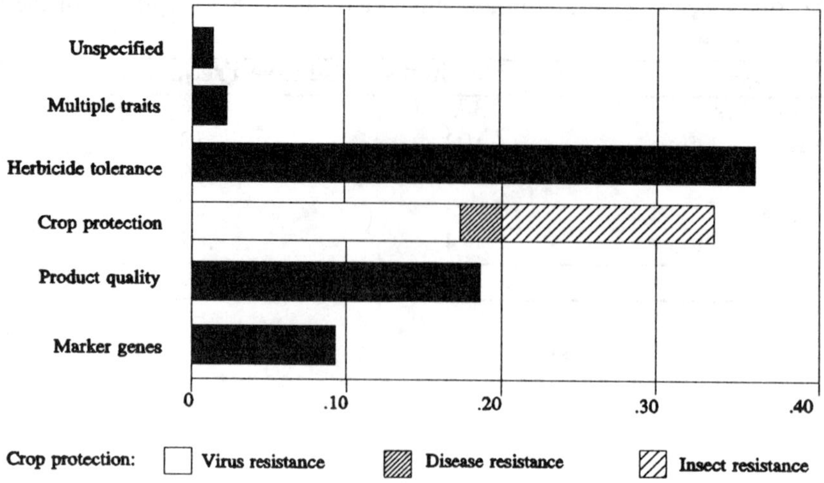

2.11

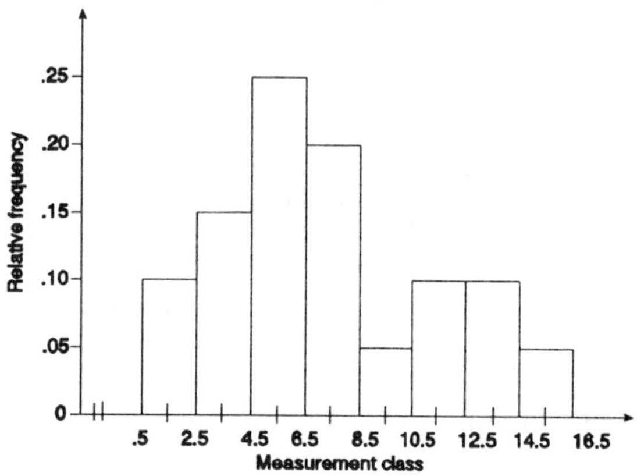

2.13 a. The original data set has 1 + 3 + 5 + 7 + 4 + 3 = 23 observations.

b. For the bottom row of the stem-and-leaf display:

The stem is 0.
The leaves are 0, 1, 2.
The numbers in the original data set are 0, 1, and 2.

c. The dot plot corresponding to all the data points is:

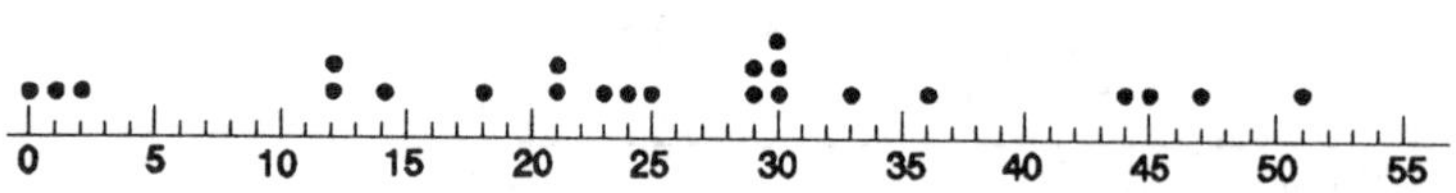

2.15 a. From the three dot plots, it appears that there is not much difference in the response rate between the Control group and the Treatment group. Both range from approximately 165 to 275. (The Treatment group did have a response of 139.) The response rate of the Familiarity group appears to be higher than for the other two groups. The response rate ranges from approximately 190 to 290.

b. First, we will construct frequency tables for each of the groups. We will use the following classes: 20.5–25.05, 25.05–30.05, 30.05–35.05, 35.05–40.05, 40.05–45.05, 45.05–50.05, and 50.05–55.05. The frequencies and relative frequencies (found by dividing the frequencies by the sample size) are listed in the following tables:

Control Group:

Measurement Class	Frequency	Relative Frequency
20.05–25.05	1	.1
25.05–30.05	6	.6
30.05–35.05	2	.2
35.05–40.05	1	.1
Totals	10	1.0

Treatment Group:

Measurement Class	Frequency	Relative Frequency
25.05–30.05	1	.1
30.05–35.05	4	.4
35.05–40.05	3	.3
40.05–45.05	2	.2
Totals	10	1.0

Familiarity Group:

Measurement Class	Frequency	Relative Frequency
35.05–40.05	1	.1
40.05–45.05	4	.4
45.05–50.05	4	.4
50.05–55.05	1	.1
Totals	10	1.0

The relative frequency histograms for the three groups are as follows:

Control Group

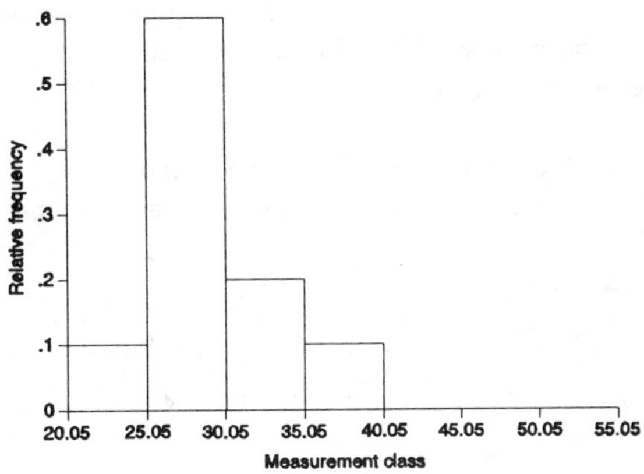

Treatment Group

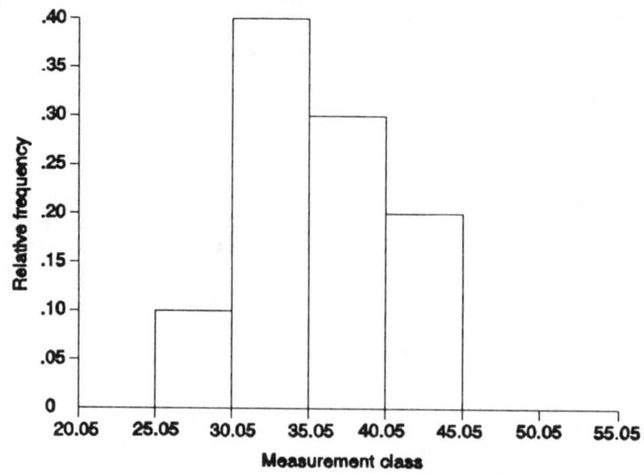

Familiarity Group

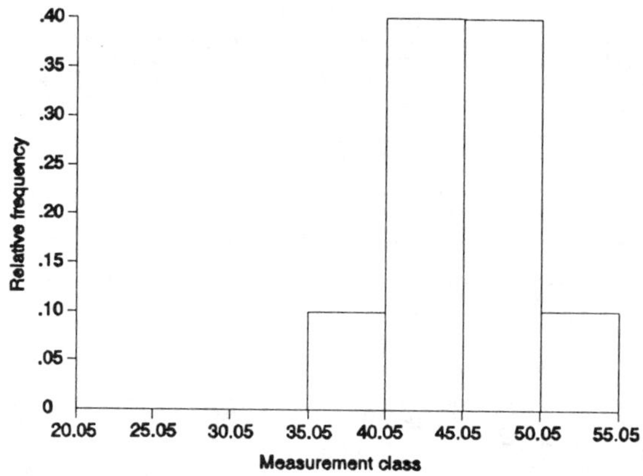

The three relative frequency histograms of the accuracy scores for the three groups indicate that the Familiarity group had the highest accuracy rate, the Treatment group had the second highest accuracy rate, and the Control group had the lowest accuracy rate.

2.17 a. The preponderance of observations are below 3000, while two observations are in the high 3000 area and five are 4000 or more. Since SPSS is using a stem which is the 1000's place rounded down to an even number, it is not always possible to determine the actual value of the five largest rates. For example, "8 . 1" could represent a number near 8100 or 9100. The leafs are rounded to the nearest 100. The largest value in the data set is 11,968.23, which is rounded to a representation of "12 . 0".

b. In the second display, the stem is the 1000's place and the leaf is the 100's place. The data seem to be clustering in the 0 to 3000 range with some extreme values.

2.19 The dot plot for these data is:

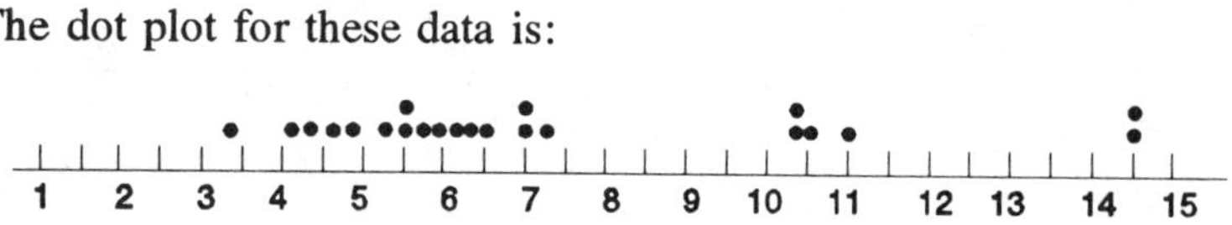

From the dot plot, most of the PMI's range from 3 to 7.5 (16 of the 22). The rest of the PMI's range from 10 to 15.

2.21 a. There are only 19 differences above 0, which indicates that only 19 states had higher scores in 1990 than in 1975. On the other hand, there are 31 differences less than 0, which indicates that 31 states had higher scores in 1975 than in 1990. This agrees with the conclusion in Exercise 2.20 that scores appear to be lower in 1990 than in 1975.

b. The largest improvement in the table is 100, which is associated with Alabama.

2.23 a. Using Minitab, the stem-and-leaf display for the data is:

```
Stem-and-Leaf of LOSS          N = 19
Leaf Unit = 1.0

  8      0 11234559
 (3)     1 123
  8      2 00
  6      3 9
  5      4 (6)
  4      5 (6)
  3      6 (1)(5)
  1      7
  1      8
  1      9
  1     10 (0)
```

b. The numbers circled on the display in part a are associated with the eclipses of Saturnian satellites.

c. Since the five largest numbers are associated with eclipses of Saturnian satellites, it is much more likely that the greater light loss is associated with eclipses rather than occults.

2.25 a. $\sum x = 5 + 1 + 3 + 2 + 1 = 12$

b. $\sum x^2 = 5^2 + 1^2 + 3^2 + 2^2 + 1^2 = 40$

c. $\sum (x - 1) = (5 - 1) + (1 - 1) + (3 - 1) + (2 - 1) + (1 - 1) = 7$

d. $\sum (x - 1)^2 = (5 - 1)^2 + (1 - 1)^2 + (3 - 1)^2 + (2 - 1)^2 + (1 - 1)^2 = 21$

e. $(\sum x)^2 = (5 + 1 + 3 + 2 + 1)^2 = 12^2 = 144$

2.27 Using the results from Exercise 2.25,

a. $\sum x^2 - \frac{(\sum x)^2}{5} = 40 - \frac{144}{5} = 40 - 28.8 = 11.2$

b. $\sum (x - 2)^2 = (5 - 2)^2 + (1 - 2)^2 + (3 - 2)^2 + (2 - 2)^2 + (1 - 2)^2 = 12$

c. $\sum x^2 - 10 = 40 - 10 = 30$

2.29 Assume the data are a sample. The mode is the observation that occurs most frequently. For this sample, the mode is 15, which occurs 3 times.

The sample mean is:

$$\bar{x} = \frac{\sum x}{n} = \frac{18 + 10 + 15 + 13 + 17 + 15 + 12 + 15 + 18 + 16 + 11}{11} = \frac{160}{11} = 14.545$$

The median is the middle number when the data are arranged in order. The data arranged in order are: 10, 11, 12, 13, 15, 15, 15, 16, 17, 18, 18. The middle number is the 6th number, which is 15.

2.31 The median is the middle number once the data have been arranged in order. If n is even, there is not a single middle number. Thus, to compute the median, we take the average of the middle two numbers. If n is odd, there is a single middle number. The median is this middle number.

A data set with 5 measurements arranged in order is 1, 3, 5, 6, 8. The median is the middle number, which is 5.

A data set with 6 measurements arranged in order is 1, 3, 5, 5, 6, 8. The median is the average of the middle two numbers which is $\frac{5 + 5}{2} = \frac{10}{2} = 5$.

2.33 a. $\bar{x} = \frac{\sum x}{n} = \frac{85}{10} = 8.5$

b. $\bar{x} = \frac{400}{16} = 25$

c. $\bar{x} = \frac{35}{45} = .78$

d. $\bar{x} = \frac{242}{18} = 13.44$

2.35 a. For a distribution that is skewed to the left, the mean is less than the median.

b. For a distribution that is skewed to the right, the mean is greater than the median.

c. For a symmetric distribution, the mean and median are equal.

2.37 a. From the printout, the mean is -0.1544 and the median is -0.11. Since these two values are very close together, both are good representatives of the middle of the data set.

b. The outlier is -1.07.

c. After deleting -1.07, the mean is:

$$\bar{x} = \frac{\sum x}{n} = \frac{(-.08) + (-.06) + (-.15) + \cdots + (-.16)}{24} = \frac{-2.79}{24} = -.11625$$

The median is the average of the middle two numbers once the data are arranged in order.

The middle two numbers are 10 and 11. The median is $\frac{(-.10) + (-.11)}{2} = -.105$

The mean changes from $-.1544$ to $-.11625$ while the median changes from $-.11$ to $-.105$. The mean changes much more than the median when the outlier is removed.

2.39 Of all the applications, there could be some with 0 letters, 1 letter, 2 letters, 3 letters, etc. Of all of the possibilities, more applications had 3 letters than had any other number (mode). Since the median was also 3, at least half of the applications had 3 or more letters. The mean was 2.28. Since this is quite a bit smaller than the median, several of the applications must have had 0 or 1 letter.

2.43 a. Since two rats did not yield an escape time before the experiment was terminated, we cannot compute the mean escape time for the ten rats involved in the experiment.

b. It is appropriate to compute the median, because the two N's represent extremely large times and have no effect on the calculation of the median. The ordered data are:

38 56 95 100 104 116 122 135 N N

Then, median $= \dfrac{104 + 116}{2} = 110$

2.45 a. The mean average SAT score in 1975 (955.3) is greater than the mean average score in 1990 (947.46). In addition, the median average SAT score in 1975 (952) is greater than the median average score in 1990 (937.5).

b. In both years, the mean is greater than the median. This implies that both data sets are skewed to the right.

2.47 a. $s^2 = \dfrac{\sum x^2 - \dfrac{(\sum x)^2}{n}}{n - 1} = \dfrac{84 - \dfrac{20^2}{10}}{10 - 1} = 4.8889$ $\qquad s = \sqrt{4.8889} = 2.211$

b. $s^2 = \dfrac{\sum x^2 - \dfrac{(\sum x)^2}{n}}{n - 1} = \dfrac{380 - \dfrac{100^2}{40}}{40 - 1} = 3.3333$ $\qquad s = \sqrt{3.3333} = 1.826$

c. $s^2 = \dfrac{\sum x^2 - \dfrac{(\sum x)^2}{n}}{n - 1} = \dfrac{18 - \dfrac{17^2}{20}}{20 - 1} = .1868$ $\qquad s = \sqrt{.1868} = .432$

2.49 a. Range $= 42 - 37 = 5$

$s^2 = \dfrac{\sum x^2 - \dfrac{(\sum x)^2}{n}}{n - 1} = \dfrac{7935 - \dfrac{199^2}{5}}{5 - 1} = 3.7$ $\qquad s = \sqrt{3.7} = 1.92$

b. Range $= 100 - 1 = 99$

$s^2 = \dfrac{\sum x^2 - \dfrac{(\sum x)^2}{n}}{n - 1} = \dfrac{25{,}795 - \dfrac{303^2}{9}}{9 - 1} = 1{,}949.25$ $\qquad s = \sqrt{1{,}949.25} = 44.15$

c. Range $= 100 - 2 = 98$

$s^2 = \dfrac{\sum x^2 - \dfrac{(\sum x)^2}{n}}{n - 1} = \dfrac{20{,}033 - \dfrac{295^2}{8}}{8 - 1} = 1{,}307.84$ $\qquad s = \sqrt{1{,}307.84} = 36.16$

2.51 This is one possibility for the two data sets.

Data Set 1: 1, 1, 2, 2, 3, 3, 4, 4, 5, 5
Data Set 2: 1, 1, 1, 1, 1, 5, 5, 5, 5, 5

$$\bar{x}_1 = \frac{\sum x}{n} = \frac{1 + 1 + 2 + 2 + 3 + 3 + 4 + 4 + 5 + 5}{10} = \frac{30}{10} = 3$$

$$\bar{x}_2 = \frac{\sum x}{n} = \frac{1 + 1 + 2 + 2 + 3 + 3 + 4 + 4 + 5 + 5}{10} = \frac{30}{10} = 3$$

Therefore, the two data sets have the same mean. The variances for the two data sets are:

$$s_1^2 = \frac{\sum x^2 - \frac{(\sum x)^2}{n}}{n - 1} = \frac{110 - \frac{30^2}{10}}{9} = \frac{20}{9} = 2.2222$$

$$s_1^2 = \frac{\sum x^2 - \frac{(\sum x)^2}{n}}{n - 1} = \frac{110 - \frac{30^2}{10}}{9} = \frac{20}{9} = 4.4444$$

The dot diagrams for the two data sets are shown below.

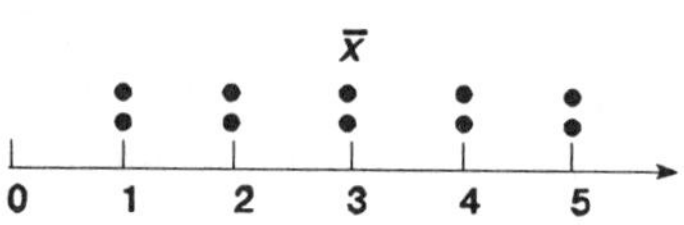

Data set 1

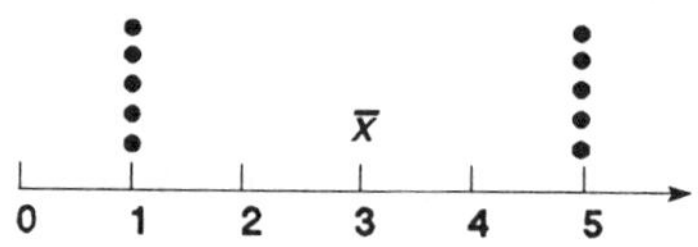

Data set 2

2.53 a. Range = 3 − 0 = 3

$$s^2 = \frac{\sum x^2 - \frac{(\sum x)^2}{n}}{n - 1} = \frac{15 - \frac{7^2}{5}}{5 - 1} = 1.3 \qquad s = \sqrt{1.3} = 1.1402$$

b. After adding 3 to each of the data points,

Range = 6 − 3 = 3

$$s^2 = \frac{\sum x^2 - \frac{(\sum x)^2}{n}}{n - 1} = \frac{102 - \frac{22^2}{5}}{5 - 1} = 1.3 \qquad s = \sqrt{1.3} = 1.1402$$

c. After subtracting 4 from each of the data points,

Range = −1 − (−4) = 3

$$s^2 = \frac{\sum x^2 - \frac{(\sum x)^2}{n}}{n - 1} = \frac{39 - \frac{(-13)^2}{5}}{5 - 1} = 1.3 \qquad s = \sqrt{1.3} = 1.1402$$

d. The range, variance, and standard deviation remain the same when any number is added to or subtracted from each measurement in the data set.

2.55 The first professor would probably be preferred because of the lower standard deviation of grade-point averages. Students in the first professor's class are less likely to have extremely low grade-point averages than are students in the second professor's class.

2.59 a. The unit of measurement of the variable of interest is dollars (the same as the mean and standard deviation). Based on this, the data are quantitative.

b. Since no information is given about the shape of the data set, we can only use Chebyshev's Rule.

$900 is 2 standard deviations below the mean, and $2100 is 2 standard deviations above the mean. Using Chebyshev's Rule, at least 3/4 of the measurements (or $3/4 \times 200 = 150$ measurements) will fall between $900 and $2100.

$600 is 3 standard deviations below the mean and $2400 is 3 standard deviations above the mean. Using Chebyshev's Rule, at least 8/9 of the measurements (or $8/9 \times 200 \approx 178$ measurements) will fall between $600 and $2400.

$1200 is 1 standard deviation below the mean and $1800 is 1 standard deviation above the mean. Using Chebyshev's Rule, nothing can be said about the number of measurements that will fall between $1200 and $1800.

$1500 is equal to the mean and $2100 is 2 standard deviations above the mean. Using Chebyshev's Rule, at least 3/4 of the measurements (or $3/4 \times 200 = 150$ measurements) will fall between $900 and $2100. It is possible that all of the 150 measurements will be between $900 and $1500. Thus, nothing can be said about the number of measurements between $1500 and $2100.

2.61 According to the Empirical Rule:

a. Approximately 68% of the measurements will be contained in the interval $\bar{x} - s$ to $\bar{x} + s$.

b. Approximately 95% of the measurements will be contained in the interval $\bar{x} - 2s$ to $\bar{x} + 2s$.

c. Essentially all the measurements will be contained in the interval $\bar{x} - 3s$ to $\bar{x} + 3s$.

2.63 Using Chebyshev's Rule, at least 8/9 of the measurements will fall within 3 standard deviations of the mean. Thus, the range of the data would be around 6 standard deviations. Using the Empirical Rule, approximately 95% of the observations are within 2 standard deviations of the mean. Thus, the range of the data would be around 4 standard deviations. We would expect the standard deviation to be somewhere between Range/6 and Range/4.

For our data, the range $= 760 - 135 = 625$.

The Range/6 $= 625/6 = 104.17$ and Range/4 $= 625/4 = 156.25$.

Therefore, I would estimate that the standard deviation of the data set is between 104.17 and 156.25.

It would not be feasible to have a standard deviation of 25. If the standard deviation were 25, the data would span $625/25 = 25$ standard deviations. This would be extremely unlikely.

2.65 a. Since the mean of the data (2.28) is smaller than the median, the data are probably skewed to the left. Also, there probably aren't too many applications with more than three letters. However, because the standard deviation is more than half of the mean, there must be a couple of very large numbers in the data set. A possible histogram of the data is:

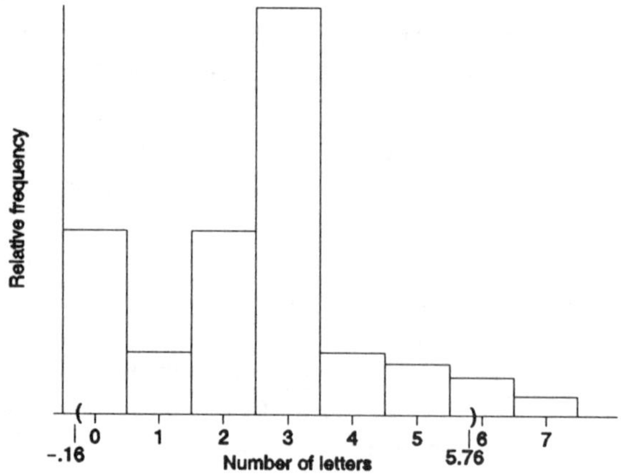

b. Since the histogram drawn is approximately mound-shaped, about 95% of the observations will fall within 2 standard deviations of the mean. This interval is:

$$\bar{x} \pm 2s \Rightarrow 2.28 \pm 2(1.48) \Rightarrow 2.28 \pm 2.96 \Rightarrow (-.68, 5.24)$$

This interval is drawn on the graph in part a.

2.67 a. For Adults:

		Interval	Percent	Frequency
$\bar{x} \pm s$	6.45 ± 2.89	(3.56, 9.34)	≈68%	46(.68) ≈ 31.3
$\bar{x} \pm 2s$	$6.45 \pm 2(2.89)$	(0.67, 12.23)	≈95%	46(.95) ≈ 43.7

For Adolescents:

		Interval	Percent	Frequency
$\bar{x} \pm s$	10.89 ± 2.48	(8.41, 13.37)	≈68%	19(.68) ≈ 12.9
$\bar{x} \pm 2s$	$10.89 \pm 2(2.48)$	(5.93, 15.85)	≈95%	19(.95) ≈ 18.1

b. See numbers in the above tables.

2.69 a. From the information given, we have $\bar{x} = 375$ and $s = 25$. From Chebyshev's Rule, we know that at least three-fourths of the measurements are within the interval:

$\bar{x} \pm 2s$, or (325, 425)

Thus, at most one-fourth of the measurements exceed 425. In other words, more than 425 vehicles used the intersection on at most 25% of the days.

b. According to the Empirical Rule, approximately 95% of the measurements are within the interval:

$\bar{x} \pm 2s$, or (325, 425)

This leaves approximately 5% of the measurements to lie outside the interval. Because of the symmetry of a mound-shaped distribution, approximately 2.5% of these will lie below 325, and the remaining 2.5% will lie above 425. Thus, on approximately 2.5% of the days, more than 425 vehicles used the intersection.

2.71 a. Since it is given that the distribution is mound-shaped, we can use the Empirical Rule. We know that 1.84% is 2 standard deviations below the mean. The Empirical Rule states that approximately 95% of the observations will fall within 2 standard deviations of the mean and, consequently, approximately 5% will lie outside that interval. Since a mound-shaped distribution is symmetric, then approximately 2.5% of the day's production of batches will fall below 1.84%.

b. If the data are actually mound-shaped, it would be extremely unusual (less than 2.5%) to observe a batch with 1.80% zinc phosphide if the true mean is 2.0%. Thus, if we did observe 1.8%, we would conclude that the mean percent of zinc phosphide in today's production is probably less than 2.0%.

2.73 a. $z = \frac{x - \bar{x}}{s} = \frac{40 - 30}{5} = 2$ (sample) 2 standard deviations above the mean.

b. $z = \frac{x - \mu}{\sigma} = \frac{90 - 89}{2} = .5$ (population) .5 standard deviations above the mean.

c. $z = \frac{x - \mu}{\sigma} = \frac{50 - 50}{5} = 0$ (population) 0 standard deviations above the mean.

d. $z = \frac{x - \bar{x}}{s} = \frac{20 - 30}{4} = -2.5$ (sample) 2.5 standard deviations below the mean.

2.75 The 50th percentile of a data set is the observation that has half of the observations less than it. Another name for the 50th percentile is the median.

2.77 a. From the problem, $\mu = 2.7$ and $\sigma = .5$

$$z = \frac{x - \mu}{\sigma} \Rightarrow z\sigma = x - \mu \Rightarrow x = \mu + z\sigma$$

For $z = 2.0$, $x = 2.7 + 2.0(.5) = 3.7$

For $z = -1.0$, $x = 2.7 - 1.0(.5) = 2.2$

For $z = .5$, $x = 2.7 + .5(.5) = 2.95$

For $z = -2.5$, $x = 2.7 - 2.5(.5) = 1.45$

b. For $z = -1.6$, $x = 2.7 - 1.6(.5) = 1.9$

c. If we assume the distribution of GPAs is approximately mound-shaped, we can use the Empirical Rule.

From the Empirical Rule, we know that $\approx .025$ or $\approx 2.5\%$ of the students will have GPAs above 3.7 (with $z = 2$). Thus, the GPA corresponding to summa cum laude (top 2.5%) will be greater than 3.7 ($z > 2$).

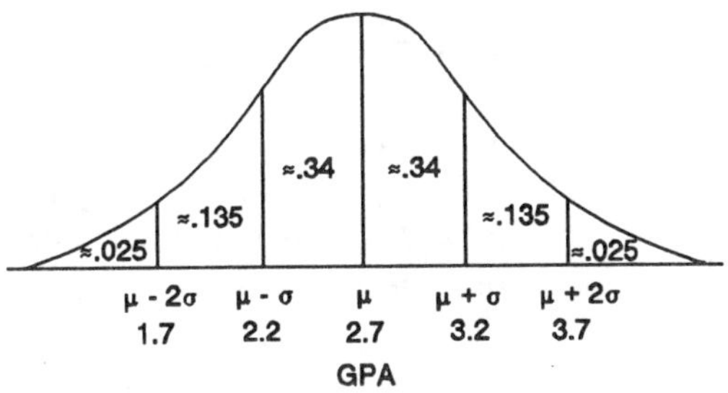

We know that $\approx .16$ or 16% of the students will have GPAs above 3.2 ($z = 1$). Thus, the limit on GPAs for cum laude (top 16%) will be greater than 3.2 ($z > 1$).

We must assume the distribution is mound-shaped.

2.79 a. If the distribution of scores was symmetric, the mean and median would be equal. The fact that the mean exceeds the median is an indication that the distribution of scores is skewed to the right.

b. It means that 90% of the scores are below 660, and 10% are above 660. (This ignores the possibility of ties, i.e., other people obtaining a score of 660.)

c. If you scored at the 94th percentile, 94% of the scores are below your score, while 6% exceed your score.

2.81 a. $z = \frac{x - \mu}{\sigma} = \frac{4 - 7}{1} = -3$

b. The z-score is low enough to suspect that the librarian's claim is incorrect. Even without any knowledge of the shape of the distribution, Chebyshev's Rule states that at least 8/9 of the measurements will fall within 3 standard deviations of the mean (and, consequently, at most 1/9 will be above $z = 3$ or below $z = -3$).

c. The Empirical Rule states that almost none of the measurements should be above $z = 3$ or below $z = -3$. Hence, the librarian's claim is even more unlikely.

d. When $\sigma = 2$, $z = \frac{x - \mu}{\sigma} = \frac{4 - 7}{1} = -1.5$

This is not an unlikely occurrence, whether or not the data are mound-shaped. Hence, we would not have reason to doubt the librarian's claim.

2.83 The 25th percentile, or lower quartile, is the measurement that has 25% of the measurements below it and 75% of the measurements above it. The 50th percentile, or median, is the measurement that has 50% of the measurements below it and 50% of the measurements above it. The 75th percentile, or upper quartile, is the measurement that has 75% of the measurements below it and 25% of the measurements above it.

2.85 a. Median is approximately 39.

b. Q_L is approximately 31.5 (Lower Quartile)

Q_U is approximately 45 (Upper Quartile)

c. IQR $= Q_U - Q_L \approx 45 - 31.5 \approx 13.5$

d. The data set is skewed to the left since the left whisker is longer.

e. 50% of the measurements are to the right of the median and 75% are to the left of the upper quartile.

2.87 a. Using Minitab, the box plot for sample A is given below.

```
                                                 --------
  O      O              *              --------I    +  I--------      *
                                                 --------
---------+---------+---------+---------+---------+--------
        100       120       140       160       180
```

Using Minitab, the box plot for sample B is given below.

```
                                 ------------
  *           ---------------I+            I--------------        *
                                 ------------
----+---------+---------+---------+---------+---------+--
   144       156       168       180       192       204
```

b. The range for sample A is larger than the range for sample B. The descriptive measures for sample A (Q_L, median, and Q_U) are all smaller than those of sample B. Both samples look somewhat symmetric (excluding outliers) since the whiskers on each box plot are the same length on both sides.

c. In sample A, the measurements 84 and 100 are outliers. These measurements fall outside the outer fences.

$$\begin{aligned} \text{Lower outer fence} &= \text{Lower hinge} - 3(\text{IQR}) \\ &\approx 158 - 3(172 - 158) \\ &= 158 - 3(14) \\ &= 116 \end{aligned}$$

In addition, 122 and 196 may be outliers. They lie outside the inner fences. In sample B, 140.4 and 206.4 may be outliers. They lie outside the inner fences.

2.89 a. Using Minitab, the box plot is:

```
          ------------
---------I     +     I-----------           *  **            *
          ------------
+---------+---------+---------+---------+---------+------
0        12        24        36        48        60
```

The median is about 18. The data appear to be skewed to the right since there are 4 suspect outliers to the right and none to the left. The variability of the data is fairly small because the IQR is fairly small, approximately 26 − 10 = 16.

b. The customers associated with the suspected outliers are customers 238, 268, 269, and 264.

c. In order to find the z-scores, we must first find the mean and standard deviation.

$$\bar{x} = \frac{\sum x}{n} = \frac{815}{40} = 20.375$$

$$s^2 = \frac{\sum x^2 - \frac{(\sum x)^2}{n}}{n - 1} = \frac{24129 - \frac{815^2}{40}}{40 - 1} = 192.90705$$

$$s = \sqrt{192.90705} = 13.89$$

The z-scores associated with the suspected outliers are:

Customer 238 $z = \dfrac{x - \bar{x}}{s} = \dfrac{47 - 20.375}{13.89} = 1.92$

Customer 268 $z = \dfrac{49 - 20.375}{13.89} = 2.06$

Customer 269 $z = \dfrac{50 - 20.375}{13.89} = 2.13$

Customer 264 $z = \dfrac{64 - 20.375}{13.89} = 3.14$

All but one of the z-scores is greater than 2. These are very unusual values.

2.91 a. The median average SAT score for 1975 ($\approx$953) is greater than the median average SAT score for 1990 ($\approx$940).

b. The interquartile range for 1990 is slightly greater than that for 1975. Therefore, the variability for 1990 is slightly greater than that for 1975. However, from the information in Exercise 2.66, the sample standard deviation for 1975 (69.5) is greater than the sample standard deviation for 1990 (64.0), which indicates the variability for 1975 is greater than that for 1990.

c. Since there are no points outside the inner fences for either year, there are no outliers.

2.93 The relative frequency histogram is:

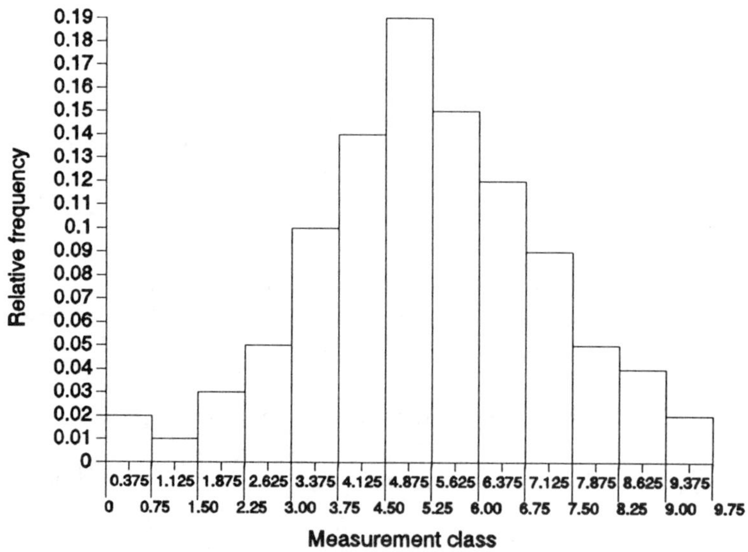

2.95 a. $z = \dfrac{x - \mu}{\sigma} = \dfrac{50 - 60}{10} = -1$

$z = \dfrac{70 - 60}{10} = 1$

$z = \dfrac{80 - 60}{10} = 2$

b. $z = \frac{x - \mu}{\sigma} = \frac{50 - 60}{5} = -2$

$z = \frac{70 - 60}{5} = 2$

$z = \frac{80 - 60}{5} = 4$

c. $z = \frac{x - \mu}{\sigma} = \frac{50 - 40}{10} = 1$

$z = \frac{70 - 40}{10} = 3$

$z = \frac{80 - 40}{10} = 4$

d. $z = \frac{x - \mu}{\sigma} = \frac{50 - 40}{100} = .1$

$z = \frac{70 - 40}{100} = .3$

$z = \frac{80 - 40}{100} = .4$

2.97 a. $s^2 = \frac{\sum x^2 - \frac{(\sum x)^2}{n}}{n - 1} = \frac{246 - \frac{63^2}{22}}{22 - 1} = 3.1234$

b. $s^2 = \frac{\sum x^2 - \frac{(\sum x)^2}{n}}{n - 1} = \frac{666 - \frac{106^2}{25}}{25 - 1} = 9.0233$

c. $s^2 = \frac{\sum x^2 - \frac{(\sum x)^2}{n}}{n - 1} = \frac{76 - \frac{11^2}{7}}{7 - 1} = 9.7857$

2.99 a. $\sum x = 4 + 6 + 6 + 5 + 6 + 7 = 34$

$\sum x^2 = 4^2 + 6^2 + 6^2 + 5^2 + 6^2 + 7^2 = 198$

$\bar{x} = \frac{\sum x}{n} = \frac{34}{6} = 5.67$

$s^2 = \frac{\sum x^2 - \frac{(\sum x)^2}{n}}{n - 1} = \frac{198 - \frac{34^2}{6}}{6 - 1} = \frac{5.3333}{5} = 1.0667$

$s = \sqrt{1.0667} = 1.03$

b. $\sum x = -1 + 4 + (-3) + 0 + (-3) + (-6) = -9$

$\sum x^2 = (-1)^2 + 4^2 + (-3)^2 + 0^2 + (-3)^2 + (-6)^2 = 71$

$\bar{x} = \frac{\sum x}{n} = \frac{-9}{6} = -\$1.5$

$$s^2 = \frac{\sum x^2 - \frac{(\sum x)^2}{n}}{n-1} = \frac{71 - \frac{(-9)^2}{6}}{6-1} = \frac{57.5}{5} = 11.5 \text{ dollars squared}$$

$$s = \sqrt{11.5} = \$3.39$$

c. $$\sum x = \frac{3}{5} + \frac{4}{5} + \frac{2}{5} + \frac{1}{5} + \frac{1}{16} = 2.0625$$

$$\sum x^2 = \left(\frac{3}{5}\right)^2 + \left(\frac{4}{5}\right)^2 + \left(\frac{2}{5}\right)^2 + \left(\frac{1}{5}\right)^2 + \left(\frac{1}{16}\right)^2 = 1.2039$$

$$\bar{x} = \frac{\sum x}{n} = \frac{2.0625}{5} = .4125\%$$

$$s^2 = \frac{\sum x^2 - \frac{(\sum x)^2}{n}}{n-1} = \frac{1.2039 - \frac{2.0625^2}{5}}{5-1} = \frac{.3531}{4} = .0883\% \text{ squared}$$

$$s = \sqrt{.0883} = .30\%$$

d. (a) Range = 7 − 4 = 3

(b) Range = \$4 − (\$−6) = \$10

(c) Range = $\frac{4}{5}\% - \frac{1}{16}\% = \frac{64}{80}\% - \frac{5}{80}\% = \frac{59}{80}\% = .7375\%$

2.101 From the pie charts, almost two-thirds of the immigrants in 1951-60 came from Canada or Europe, while in 1981-90, only 15% came from the same region. Also, in 1951-60, only 6% of the immigrants came from Asia, while in 1981-90, 45% came from Asia. The percentage of immigrants from the other regions stayed fairly constant over the two time periods.

2.103 a. Since the mean is greater than the median, the distribution of the radiation levels is skewed to the right.

b. $\bar{x} \pm s \Rightarrow 10 \pm 3 \Rightarrow (7, 13)$; $\bar{x} \pm 2s \Rightarrow 10 \pm 2(3) \Rightarrow (4, 16)$; $\bar{x} \pm 3s \Rightarrow 10 \pm 3(3) \Rightarrow (1, 19)$

Interval	Chebyshev's	Empirical
(7, 13)	At least 0	≈68%
(4, 16)	At least 75%	≈95%
(1, 19)	At least 88.9%	≈100%

Since the data are skewed to the right, Chebyshev's Rule is probably more appropriate in this case.

c. The background level is 4. Using Chebyshev's Rule, at least 75% or .75(50) ≈ 38 homes are above the background level. Using the Empirical Rule, ≈ 97.5% or .975(50) ≈ 49 homes are above the background level.

d. $$z = \frac{x - \bar{x}}{s} = \frac{20 - 10}{3} = 3.333$$

It is unlikely that this new measurement came from the same distribution as the other 50. Using either Chebyshev's Rule or the Empirical Rule, it is very unlikely to see any observations more than 3 standard deviations from the mean.

2.105 a. Using Minitab, a stem-and-leaf display of the data is:

```
Stem-and-leaf of VELOCITY        N = 51
Leaf Unit = 100

 1    18 4
 3    18 79
12    19 001112444
18    19 566788
20    20 12
21    20 7
21    21
23    21 99
(5)   22 11344
23    22 5666777777889
10    23 001222344
 1    23
 1    24
 1    24 9
```

b. From this stem-and-leaf display, it is fairly obvious that there are two different distributions since there are two groups of data.

c. Since there appears to be two distributions, we will compute two sets of numerical descriptive measures. We will call the group with the smaller velocities A1775A and the group with the larger velocities A1775B.

For A1775A:

$$\bar{x} = \frac{\sum x}{n} = \frac{408{,}707}{21} = 19{,}462.2$$

$$s^2 = \frac{\sum x^2 - \frac{\left(\sum x\right)^2}{n}}{n - 1} = \frac{7{,}960{,}019{,}531 - \frac{408{,}707^2}{21}}{21 - 1} = 283{,}329.3$$

$$s = \sqrt{283{,}329.3} = 532.29$$

For A1775B:

$$\bar{x} = \frac{\sum x}{n} = \frac{685{,}154}{30} = 22{,}838.5$$

$$s^2 = \frac{\sum x^2 - \frac{\left(\sum x\right)^2}{n}}{n - 1} = \frac{15{,}656{,}992{,}942 - \frac{685{,}154^2}{30}}{30 - 1} = 314{,}694.83$$

$$s = \sqrt{314{,}694.83} = 560.98$$

d. To determine which of the two clusters this observation probably belongs to, we will compute z-scores for this observation for each of the two clusters.

For A1775A:

$$z = \frac{20{,}000 - 19{,}462.2}{532.29} = 1.01$$

Since this z-score is so small, it would not be unlikely that this observation came from this cluster.

For A1775B:

$$z = \frac{20{,}000 - 22{,}838.5}{560.98} = -5.06$$

Since this z-score is so large (in magnitude), it would be very unlikely that this observation came from this cluster.

Thus, this observation probably came from the cluster A1775A.

2.107 a. The number of tornadoes per year is a number; therefore, it is quantitative. The month of peak occurrence is a category; therefore, it is qualitative.

b. A bar chart is used to display the month of peak tornado occurrence in the 50 states:

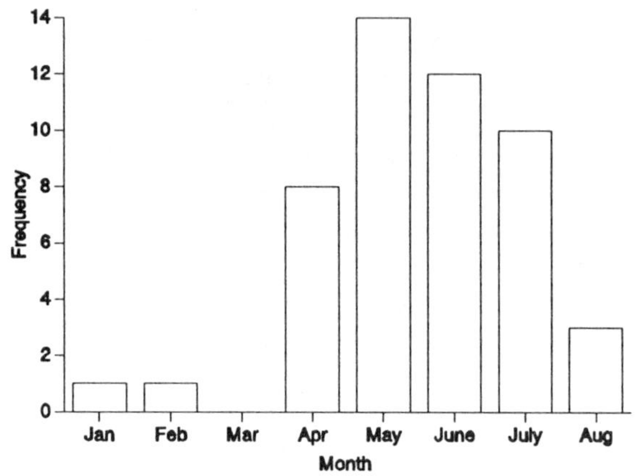

2.109 a. A stem-and-leaf display generated by Minitab is given below. It indicates a fairly symmetric, slightly bimodal distribution.

```
   3   0 678
  11   1 02344444
 (5)   1 55668
   9   2 011122
   3   2 568
```

b. $\bar{x} = \sum x = \frac{21 + \cdots + 10}{25} = \frac{412}{25} = 16.48$

Median = 15 (13th number, after ordering)

Mode = 14

Using Minitab, the frequency histogram of the data is as follows. To construct a relative frequency histogram, divide the frequency by 25.

```
                 Histogram of C1      N = 25

                 Midpoint      Count
                        6          1  *
                        8          2  **
                       10          1  *
                       12          1  *
       Mode  →         14          6  ******
Median ----------→
       Mean  →         16          4  ****
                       18          1  *
                       20          1  *
                       22          5  *****
                       24          0
                       26          2  **
                       28          1  *
```

c. Range = 28 − 6 = 22

$$s = \frac{\text{Range}}{4} = \frac{22}{4} = 5.5$$

d. $$s^2 = \frac{\sum x^2 - \frac{(\sum x)^2}{n}}{n - 1} = \frac{7604 - \frac{412^2}{25}}{25 - 1} = 33.9267$$

$$s = \sqrt{s^2} = \sqrt{33.9267} = 5.82$$

This agrees with the approximation of part c.

e. $\bar{x} \pm 2s \Rightarrow 16.48 \pm 2(5.82) \Rightarrow (4.84, 28.12)$

All of the data are in this range.

2.111 We first evaluate how extreme a value of 200 ppm for a PCB count is by converting this amount to a z-score.

$$z = \frac{200 - 25}{5} = \frac{175}{5} = 35$$

By using Chebyshev's Rule, the proportion of times a PCB count of 200 should occur is **at most** $1/35^2 = 1/1225 \approx .0008$.

With such a small probability for the maximum probability of observing this value, it would definitely be considered to be an outlier.

Probability

Chapter

3

3.1 a. Since the probabilities must sum to 1,

$$P(E_3) = 1 - P(E_1) - P(E_2) - P(E_4) - P(E_5) = 1 - .1 - .2 - .1 - .1 = .5$$

b. $P(E_3) = 1 - P(E_3) - P(E_2) - P(E_4) - P(E_5)$
$\Rightarrow 2P(E_3) = 1 - .1 - .2 - .1 \Rightarrow 2P(E_3) = .6 \Rightarrow P(E_3) = .3$

c. $P(E_3) = 1 - P(E_1) - P(E_2) - P(E_4) - P(E_5) = 1 - .1 - .1 - .1 - .1 = .6$

3.3 $P(A) = P(1) + P(2) + P(3) = .05 + .20 + .30 = .55$
$P(B) = P(1) + P(3) + P(5) = .05 + .30 + .15 = .50$
$P(C) = P(1) + P(2) + P(3) + P(5) = .05 + .20 + .30 + .15 = .70$

3.5 a. $P(A) = P(2) + P(3) + P(5) = 2/9 + 1/9 + 1/9 = 4/9$

b. $P(B) = P(4) + P(6) = 2/9 + 1/9 = 3/9 = 1/3$

c. Since there are no events that occur in both A and B at the same time, $P(A \text{ and } B \text{ simultaneously}) = 0$.

3.7 The simple events are:

1, H	1, T
2, H	2, T
3, H	3, T
4, H	4, T
5, H	5, T
6, H	6, T

Since each event is equally likely if the die and coin are fair, each event will have a probability of 1/12.

$P(A) = P(6, H) = 1/12$

$P(B) = P(2, T) + P(4, T) + P(6, T) = 1/12 + 1/12 + 1/12 = 3/12 = 1/4$

$P(C) = P(2, T) + P(4, T) + \text{P}(6, T) + P(2, H) + P(4, H) + P(6, H)$
$= 1/12 + 1/12 + 1/12 + 1/12 + 1/12 + 1/12 = 6/12 = 1/2$

$P(D) = P(1, T) + P(2, T) + P(3, T) + P(4, T) + P(5, T) + P(6, T)$
$= 1/12 + 1/12 + 1/12 + 1/12 + 1/12 + 1/12 = 6/12 = 1/2$

3.9 Each student will obtain slightly different proportions. However, the proportions should be close to $P(A) = 1/10$, $P(B) = 6/10$ and $P(C) = 3/10$.

3.11 a. The sample points would correspond to the possible outcomes: Violence to others, Sympathetic, Harm to self, Comic images, and Criticism of definitions.

b. From the sample, the number of times each of these sample points occurred was different. We would expect the probability of each sample point to be close to the relative frequency of that outcome.

c. Reasonable probabilities would be the relative frequencies of the sample points:

Media Coverage	Frequency	Relative Frequency
Violence to others	373	373/562 = .6637
Sympathetic	102	102/562 = .1815
Harm to self	71	71/562 = .1263
Comic images	12	12/562 = .0214
Criticism of definitions	4	4/562 = .0071
Total	562	1.0000

d. This probability would be .1815 + .0214 = .2029.

3.13 a. From the problem, scientists found that 6 of the 21 families studied had mutations in the S182 gene on chromosome 14. Thus, a good estimate of this probability is 6/21 = .286.

b. The reliability of this estimate is probably not very high since only 21 families were sampled. To get a more reliable estimate, a larger set of families would be needed.

3.15 If each of 3 players uses a fair coin, the sample space is:

HHH	*HTT*
HHT	*THT*
HTH	*TTH*
THH	*TTT*

Since each event is equally likely, each event will have probability 1/8.

The probability of odd man out on the first roll is:

$$P(HHT) + P(HTH) + P(THH) + P(HTT) + P(THT) + P(TTH)$$
$$= 1/8 + 1/8 + 1/8 + 1/8 + 1/8 + 1/8 = 6/8 = 3/4$$

If one player uses a two-headed coin, the sample space will be (assume 3rd player has the two-headed coin):

HHH	*THH*
HTH	*TTH*

Since each event is equally likely, each will have probability 1/4.

The probability the 3rd player is odd man out is $P(TTH) = 1/4$.

From the first part, the probability the 3rd player is odd man out is $P(HHT) + P(TTH) = 1/8 + 1/8 = 1/4$. Thus, the two probabilities are the same.

3.17 a. The odds in favor of Snow Chief are $\frac{1}{3}$ to $\left(1 - \frac{1}{3}\right)$ or $\frac{1}{3}$ to $\frac{2}{3}$ or 1 to 2

b. If the odds are 1 to 1, $P(\text{Snow Chief will win}) = \frac{1}{1 + 1} = \frac{1}{2} = .5$

c. If the odds against Snow Chief winning are 3 to 2, the odds for Snow Chief winning are 2 to 3. The probability Snow Chief will win is $\frac{2}{2 + 3} = \frac{2}{5} = .4$

3.19 Let event A be the event that a person with a newly earned Ph.D. in science or engineering will obtain one of the available jobs. Then $P(A) = 1000/1250 = .8$.

3.21 Two events are mutually exclusive if they have no simple events in common. A possible Venn diagram of two mutually exclusive events is:

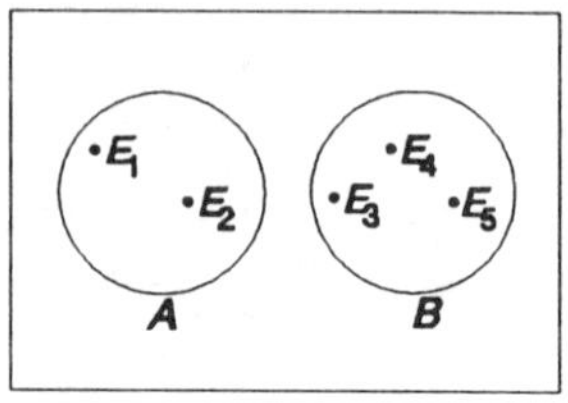

3.23 a. $P(A) = P(E_1) + P(E_2) + P(E_3) + P(E_5) + P(E_6) = \frac{1}{5} + \frac{1}{5} + \frac{1}{5} + \frac{1}{20} + \frac{1}{10} = \frac{15}{20} = \frac{3}{4}$

b. $P(B) = P(E_2) + P(E_3) + P(E_4) + P(E_7) = \frac{1}{5} + \frac{1}{5} + \frac{1}{20} + \frac{1}{5} = \frac{13}{20}$

c. $P(A \cup B) = P(E_1) + P(E_2) + P(E_3) + P(E_4) + P(E_5) + P(E_6) + P(E_7)$

$= \frac{1}{5} + \frac{1}{5} + \frac{1}{5} + \frac{1}{20} + \frac{1}{20} + \frac{1}{10} + \frac{1}{5} = 1$

d. $P(A \cap B) = P(E_2) + P(E_3) = \frac{1}{5} + \frac{1}{5} = \frac{2}{5}$

e. $P(A') = 1 - P(A) = 1 - \frac{3}{4} = \frac{1}{4}$

f. $P(B') = 1 - P(B) = 1 - \frac{13}{20} = \frac{7}{20}$

g. $P(A \cup A') = P(E_1) + P(E_2) + P(E_3) + P(E_4) + P(E_5) + P(E_6) + P(E_7)$

$= \frac{1}{5} + \frac{1}{5} + \frac{1}{5} + \frac{1}{20} + \frac{1}{20} + \frac{1}{10} + \frac{1}{5} = 1$

h. $P(A' \cap B) = P(E_4) + P(E_7) = \frac{1}{20} + \frac{1}{5} = \frac{5}{20} = \frac{1}{4}$

3.25 The relative frequencies found by dividing the frequencies by $n = 8000$, are given in table form.

		Income		
		(D) Under \$20,000	(E) \$20,000–\$50,000	(F) Over \$50,000
Age	(A) Under 25	$\frac{950}{8000}$	$\frac{1000}{8000}$	$\frac{50}{8000}$
	(B) 25–45	$\frac{450}{8000}$	$\frac{2050}{8000}$	$\frac{1500}{8000}$
	(C) Over 45	$\frac{50}{8000}$	$\frac{950}{8000}$	$\frac{1000}{8000}$

a. $P(B) = \frac{450}{8000} + \frac{2050}{8000} + \frac{1500}{8000} = \frac{4000}{8000} = .5$

b. $P(F) = \frac{50}{8000} + \frac{1500}{8000} + \frac{1000}{8000} = \frac{2550}{8000} = .31875$

c. $P(C \cap F) = \frac{1000}{8000} = .125$

d. $P(B \cup C) = \frac{450}{8000} + \frac{2050}{8000} + \frac{1500}{8000} + \frac{50}{8000} + \frac{950}{8000} + \frac{1000}{8000} = \frac{6000}{8000} = .75$

e. $P(A') = 1 - P(A) = 1 - \left[\frac{950}{8000} + \frac{1000}{8000} + \frac{50}{8000}\right] = 1 - \frac{2000}{8000} = \frac{6000}{8000} = .75$

f. $P(A' \cap F) = \frac{1500}{8000} + \frac{1000}{8000} = \frac{2500}{8000} = .3125$

g. $P(A \cap B) = 0$. Thus, A and B are mutually exclusive.
$P(A \cap C) = 0$. Thus, A and C are mutually exclusive.
$P(A \cap D) = 950/8000$. Thus, A and D are not mutually exclusive.
$P(A \cap E) = 1000/8000$. Thus, A and E are not mutually exclusive.
$P(A \cap F) = 50/8000$. Thus, A and F are not mutually exclusive.
$P(B \cap C) = 0$. Thus, B and C are mutually exclusive.
$P(B \cap D) = 450/8000$. Thus, B and D are not mutually exclusive.
$P(B \cap E) = 2050/8000$. Thus, B and E are not mutually exclusive.
$P(B \cap F) = 1500/8000$. Thus, B and F are not mutually exclusive.
$P(C \cap D) = 50/8000$. Thus, C and D are not mutually exclusive.
$P(C \cap E) = 950/8000$. Thus, C and E are not mutually exclusive.
$P(C \cap F) = 1000/8000$. Thus, C and F are not mutually exclusive.
$P(D \cap E) = 0$. Thus, D and E are mutually exclusive.
$P(D \cap F) = 0$. Thus, D and F are mutually exclusive.
$P(E \cap F) = 0$. Thus, E and F are mutually exclusive.

3.27 Define the following events:

E_1: {3 heads}
E_2: {2 heads}
E_3: {1 heads}
E_4: {0 heads}

a. $A = E_1 \cup E_2 \cup E_3$

$P(A) = P(E_1) + P(E_2) + P(E_3) = \frac{1}{8} + \frac{3}{8} + \frac{3}{8} = \frac{7}{8}$

b. $A = E_4'$ $\quad P(A) = 1 - P(E_4) = 1 - \frac{1}{8} = \frac{7}{8}$

3.29 a. The student is male and a binge drinker would be the event $B \cap A$.

b. The student is not a binge drinker would be the event A'.

c. The student is male or lives in a coed dorm is the event $B \cup C$.

d. The student is female and not a binge drinker would be the event $B' \cap A'$.

3.31 a. There are a total of $3 \times 4 = 12$ sample points. These sample points are:

(Hearing Aid Experience, Daily Use)
(Less than 1 yr., Less than 1)
(1–10 yrs., Less than 1)
(More than 10 yrs., Less than 1)
(Less than 1 yr., 1–4)
(1–10 yrs., 1–4)
(More than 10 yrs., 1–4)

(Hearing Aid Experience, Daily Use)
(Less than 1 yr., 4–8)
(1–10 yrs., 4–8)
(More than 10 yrs., 4–8)
(Less than 1 yr., 8–16)
(1–10 yrs., 8–16)
(More than 10 yrs., 8–16)

b. The set of all sample points is the sample space.

c. $P(C) = .00 + .03 + .06 + .12 = .21$ (Note: convert the percents to proportions.)

d. $P(G) = .15 + .28 + .12 = .55$

e. $P(A) = .04 + .06 + .07 + .15 = .32$

f. $P(D) = .04 + .04 + .00 = .08$

g. $P(E) = .06 + .09 + .03 = .18$

3.33 a. Yes, the probabilities in the table sum to 1.

$.05 + .16 + .05 + .19 + .32 + .05 + .11 + .05 + .02 = 1$

b. Let a sample point be the pair (Motivation, Talent).

A: $\{(H, H), (H, M), (H, L)\}$, $P(A) = .05 + .16 + .05 = .26$
B: $\{(H, H), (M, H), (L, H)\}$, $P(B) = .05 + .19 + .11 = .35$
C: $\{(H, H), (H, M), (M, H), (M, M)\}$, $P(C) = .05 + .16 + .19 + .32 = .72$
D: $\{(H, L), (M, L), (L, L), (L, M), (L, H)\}$, $P(D) = .05 + .05 + .11 + .05 + .02 = .28$
E: $\{(H, H)\}$, $P(E) = .05$

c. $P(A \cup B) = .05 + .16 + .05 + .19 + .11 = .56$
$P(A \cap B) = .05$
$P(A \cup C) = .05 + .16 + .05 + .19 + .32 = .77$

d. $P(A') = 1 - P(A) = 1 - .26 = .74$

The probability that a managerial prospect is not highly motivated is .74. Only about 1/4 of the prospects are highly motivated.

e. $P(A \cap B) = .05$. Thus, A and B are not mutually exclusive.
$P(A \cap C) = .21$. Thus, A and C are not mutually exclusive.
$P(A \cap D) = .05$. Thus, A and D are not mutually exclusive.
$P(A \cap E) = .05$. Thus, A and E are not mutually exclusive.
$P(B \cap C) = .24$. Thus, B and C are not mutually exclusive.
$P(B \cap D) = .11$. Thus, B and D are not mutually exclusive.
$P(B \cap E) = .05$. Thus, B and E are not mutually exclusive.
$P(C \cap D) = 0$. Thus, C and D are mutually exclusive.
$P(C \cap E) = .05$. Thus, C and E are not mutually exclusive.
$P(D \cap E) = 0$. Thus, D and E are mutually exclusive.

3.35 $P(A \mid B) = \frac{.15}{.60} = .25$ $\qquad P(B \mid A) = \frac{.15}{.30} = .5$

3.37 a.

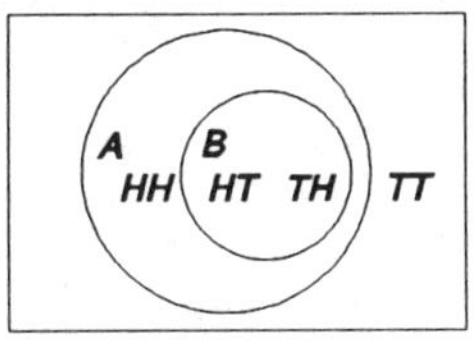

Each simple event has a probability of $\frac{1}{4}$.

b. $P(A) = P(HH) + P(HT) + P(TH) = 3\left(\frac{1}{4}\right) = \frac{3}{4}$

$P(B) = P(HT) + P(TH) = 2\left(\frac{1}{4}\right) = \frac{1}{2}$

$P(A \cap B) = P(B) = \frac{1}{2}$

c. $P(A \mid B) = \frac{1/2}{1/2} = 1$ $\qquad P(B \mid A) = \frac{1/2}{3/4} = \frac{2}{3}$

3.39 Define the following events:

A: {Crime involved white attacker and white victim}
B: {Crime involved serious injuries}
C: {Crime involved fatalities}
D: {Crime involved no injury}
E: {Crime involved non-white attacker and white victim}

a. $P(A) = \frac{6,521}{11,717} = .557$

b. $P(B) = \frac{881}{11,717} = .075$

c. $P(C \mid A) = \dfrac{P(A \cap C)}{P(A)} = \dfrac{183/11{,}717}{6{,}521/11{,}717} = \dfrac{183}{6{,}521} = .028$

d. $P(E \mid D) = \dfrac{P(E \cap D)}{P(D)} = \dfrac{1{,}801/11{,}717}{3{,}573/11{,}717} = \dfrac{1{,}801}{3{,}573} = .504$

3.41 a. Define the following events:

W: {Player wins the game Go}
F: {Player plays first (black stones)}

$P(W \cap F) = 319/577 = .553$

b. $P(W \cap F \mid CA) = 34/34 = 1$
$P(W \cap F \mid CB) = 69/79 = .873$
$P(W \cap F \mid CC) = 66/118 = .559$
$P(W \cap F \mid BA) = 40/54 = .741$
$P(W \cap F \mid BB) = 52/95 = .547$
$P(W \cap F \mid BC) = 27/79 = .342$
$P(W \cap F \mid AA) = 15/28 = .536$
$P(W \cap F \mid AB) = 11/51 = .216$
$P(W \cap F \mid AC) = 3/39 = .077$

c. There are three combinations where the player with the black stones (first) is ranked higher than the player with the white stones: *CA*, *CB*, and *BA*.

$P(W \cap F \mid CA \cup CB \cup BA) = (34 + 69 + 40)/(34 + 79 + 54) = 143/167 = .856$

d. There are three combinations where the players are of the same level: *CC*, *BB*, and *AA*.

$P(W \cap F \mid CC \cup BB \cup AA) = (66 + 52 + 15)/(118 + 95 + 28) = 133/241 = .552$

3.43 a. We will define the events the same as in Exercise 3.16.

There are a total of $9 \times 2 = 18$ sample points for this experiment. There are 9 sources of CO poisoning, and each source of poisoning has 2 possible outcomes, fatal or nonfatal. Suppose we introduce some notation to make it easier to write down the sample points. Let FI = Fire, AU = Auto exhaust, FU = Furnace, K = Kerosene or spaceheater, AP = Appliance, OG = Other gas-powered motors, FP = Fireplace, O = Other, and U = Unknown. Also, let F = Fatal and N = Nonfatal. The 18 sample points are:

FI, F	*AU, F*	*FU, F*	*K, F*	*AP, F*	*OG, F*	*FP, F*	*O, F*	*U, F*
FI, N	*AU, N*	*FU, N*	*K, N*	*AP, N*	*OG, N*	*FP, N*	*O, N*	*U, N*

$P(F \mid FI) = 63/116 = .543$

b. $P(AU \mid N) = 178/807 = .221$

c. $P(U \mid F) = 9/174 = .052$

d. The event "not fire or fireplace" would include *AU, FU, K, AP, OG, O,* and *U*.

$P(AU \cup FU \cup K \cup AP \cup OG \cup O \cup U \mid N)$
$= (178 + 345 + 18 + 63 + 73 + 19 + 42)/807 = 738/807 = .914$

3.45 Define the following events:

A: {Person chosen is black}
B: {Person chosen is PGM type 1-1}

a. $P(A) = \dfrac{6.7\% + 3.4\% + .4\%}{100\%} = .105$

b. $P(B \mid A) = \dfrac{P(A \cap B)}{P(A)} = \dfrac{\frac{6.7\%}{100\%}}{.105} = .638$

c. $P(B \mid A') = \dfrac{P(A' \cap B)}{P(A')} = \dfrac{\frac{46.3\%}{100\%}}{1 - .105} = \dfrac{.463}{.895} = .517$

3.47 a. $P(A) = P(E_1) + P(E_3) = .22 + .15 = .37$

b. $P(B) = P(E_2) + P(E_3) + P(E_4) = .31 + .15 + .22 = .68$

c. $P(A \cap B) = P(E_3) = .15$

d. $P(A \mid B) = \dfrac{P(A \cap B)}{P(B)} = \dfrac{.15}{.68} = .221$

e. $P(B \cap C) = 0$

f. $P(C \mid B) = \dfrac{P(B \cap C)}{P(B)} = \dfrac{0}{.68} = 0$

g. $P(A)P(B) = .37(.68) = .2516 \neq P(A \cap B) = .15$. Thus, *A* and *B* are not independent.

$P(A)P(C) = .37(.32) = .1184 \neq P(A \cap C) = .22$. Thus, *A* and *C* are not independent.

$P(B)P(C) = .68(.32) = .2176 \neq P(B \cap C) = 0$. Thus, *B* and *C* are not independent.

3.49 a. $P(A \cap C) = 0 \Rightarrow$ *A* and *C* are mutually exclusive.
$P(B \cap C) = 0 \Rightarrow$ *B* and *C* are mutually exclusive.

b. $P(A) = P(1) + P(2) + P(3) = .20 + .05 + .30 = .55$
$P(B) = P(3) + P(4) = .30 + .10 = .40$
$P(C) = P(5) + P(6) = .10 + .25 = .35$
$P(A \cap B) = P(3) = .30$
$P(A \mid B) = \dfrac{P(A \cap B)}{P(B)} = \dfrac{.30}{.40} = .75$

A and *B* are independent if $P(A \mid B) = P(A)$. Since $P(A \mid B) = .75$ and $P(A) = .55$, *A* and *B* are not independent.

Since A and C are mutually exclusive, they are not independent. Similarly, since B and C are mutually exclusive, they are not independent.

c. Using the probabilities of simple events,

$$P(A \cup B) = P(1) + P(2) + P(3) + P(4) = .20 + .05 + .30 + .10 = .65$$

Using the additive rule,

$$P(A \cup B) = P(A) + P(B) - P(A \cap B) = .55 + .40 - .30 = .65$$

Using the probabilities of simple events,

$$P(A \cup C) = P(1) + P(2) + P(3) + P(5) + P(6)$$
$$= .20 + .05 + .30 + .10 + .25 = .90$$

Using the additive rule,

$$P(A \cup C) = P(A) + P(C) - P(A \cap C) = .55 + .35 - 0 = .90$$

3.51 a. $P(A \cap B) = P(A) \cdot P(B) = (.4)(.2) = .08$
$P(A \mid B) = P(A) = .4$
$P(A \cup B) = P(A) + P(B) - P(A \cap B) = .4 + .2 - .08 = .52$

b. $P(A \cap B) = P(A \mid B) \cdot P(B) = (.6)(.2) = .12$

$$P(B \mid A) = \frac{P(A \cap B)}{P(A)} = \frac{.12}{.40} = .30$$

3.53 a. The simple events are:

$A \cap B$: {Microchip lasts through first use and end of first year}
$A \cap B'$: {Microchip lasts through first use and fails during first year}
A': {Microchip does not last through first use}

b. We are given $P(A') = .10$ and $P(B \mid A) = .99$

$P(A \cap B) = P(B \mid A)P(A) = .99(.9) = .891$ $\qquad P(A) = 1 - P(A') = 1 - .1 = .9$

Since the sum of the probabilities associated with all the simple events is 1,

$$P(A \cap B') = 1 - P(A \cap B) - P(A') = 1 - .891 - .10 = .009$$

c. $P(A \cap B) = .891$

3.55 a. The probability that all four male professional golfers make a hole in one on the same hole is:

$$(1/2780)^4 \approx 1.674 \times 10^{-14}$$

We must assume that the outcomes of all golfers are independent of each other.

b. The probability that none of the four male professional golfers make a hole-in-one on the same hole is:

$$(2779/2780)^4 \approx .9986$$

c. The probability that at least one of the four male professional golfers makes a hole-in-one on the same hole is:

$$1 - (2779/2780)^4 \approx 1 - .9986 = .0014$$

d. The probability that all four female professional golfers make a hole-in-one on the same hole is:

$$(1/2920)^4 \approx 1.376 \times 10^{-14}$$

We must assume that the outcomes of all golfers are independent of each other.

The probability that none of the four female professional golfers make a hole-in-one on the same hole is:

$$(2919/2920)^4 \approx .9986$$

The probability that at least one of the four female professional golfers makes a hole-in-one on the same hole is:

$$1 - (2919/2920)^4 \approx 1 - .9986 = .0014$$

3.57 Let us define the following events:

S: {School is a subscriber}
N: {School never uses the CCN broadcast}
F: {School uses the CCN broadcast more than 5 times per week}

From the problem, $P(S) = .40$, $P(N \mid S) = .05$, and $P(F \mid S) = .20$.

a. $P(S \cap N) = P(N \mid S)\, P(S) = .05(.40) = .02$

b. $P(S \cap F) = P(F \mid S)\, P(S) = .20(.40) = .08$

3.59 a. Refer to the simple events listed in Exercise 3.22.

$$P(\text{win on first roll}) = P(7 \text{ or } 11) = \frac{8}{36} = \frac{2}{9}$$

b. $$P(\text{lose on first roll}) = P(\text{sum of } 2) = \frac{1}{36}$$

c. If the player rolls a 3 on the first roll, the game will end on the next roll if:

1) the player rolls a 3
2) the player rolls a 7 or an 11
3) the player rolls a 2

This probability is the union of three mutually exclusive events.

$$P(\text{game ends on a second roll}) = \frac{2}{36} + \frac{8}{36} + \frac{1}{36} = \frac{11}{36}$$

3.61 Define the following events.

D: {Carrier's child acquires the disease}
N: {Carrier's child does not acquire the disease}

A simple event is a triplet consisting of the conditions of the carrier's three children. Note that $P(D) = 1/4$ and $P(N) = 1 - P(D) = 3/4$.

a. $P(NNN) = P(N \cap N \cap N) = P(N)P(N)P(N) = (3/4)^3 = 27/64$, since transmission of the disease to one child is independent of transmission to another.

b. $P(DDD) = P(D \cap D \cap D) = \{P(D)\}^3 = (1/4)^3 = 1/64$

c.
$$\begin{aligned} P(\text{exactly one diseased}) &= P(NND) + P(NDN) + P(DNN) \\ &= P(N \cap N \cap D) + P(N \cap D \cap N) + P(D \cap N \cap N) \\ &= 3P(N)P(N)P(D) \\ &= 3(1/4)(3/4)^2 = 27/64 \end{aligned}$$

3.63 a. Let the 6 elements be A, B, C, D, E, F.

The number of ways to draw 3 elements is:

ABC	ACE	BCD	BEF
ABD	ACF	BCE	CDE
ABE	ADE	BCF	CDF
ABF	ADF	BDE	CEF
ACD	AEF	BDF	DEF

Using combinatorial mathematics, the number of ways to select a sample of $n = 3$ elements is:

$$\binom{6}{3} = \frac{6!}{3!(6-3)!} = \frac{6 \cdot 5 \cdot 4 \cdot 3 \cdot 2 \cdot 1}{3 \cdot 2 \cdot 1 \cdot 3 \cdot 2 \cdot 1} = 20$$

b. Since there are 20 possibilities, the probability of selecting any particular sample is 1/20.

3.65 First, number the intersections from 1 to 5000. Using the random number table, select a starting point that contains 4 digits. Following either the row or column, select successive 4 digit numbers until 50 different 4 digit numbers between 0001 and 5000 are selected. The intersections corresponding to these 50 four digit numbers are selected for sampling.

The second method requires the rows to be numbered from 00 to 99 and the columns to be numbered from 1 to 50. Using the random number table, select a starting point that contains 2 digits. Following either the row or column, select successive 2 digit numbers in sets of 2. The first 2 digit number will correspond to the row number. The second 2 digit number must be between 1 and 50 and corresponds to the column number. This procedure is followed until 50 pairs of 2 digit numbers are selected that correspond to 50 different intersections.

3.67 First, number the households from 1 to 534,322. Using the random number table, select a starting point that consists of 6 digits. Following either the row or column, select successive 6 digit numbers between 1 and 534,322 until 1000 different 6 digit numbers have been selected. The sample will consist of the 1000 households corresponding to the 1000 different numbers.

3.69 a. $P(A \cap B) = 0$

$P(B \cap C) = P(2) = .2$

$P(A \cup C) = P(1) + P(2) + P(3) + P(5) + P(6) = .3 + .2 + .1 + .1 + .2 = .9$

$$\begin{aligned} P(A \cup B \cup C) &= P(1) + P(2) + P(3) + P(4) + P(5) + P(6) \\ &= .3 + .2 + .1 + .1 + .1 + .2 = 1 \end{aligned}$$

$P(B') = P(1) + P(3) + P(5) + P(6) = .3 + .1 + .1 + .2 = .7$

$P(A' \cap B) = P(2) + P(4) = .2 + .1 = .3$

$$P(B \mid C) = \frac{P(B \cap C)}{P(C)} = \frac{P(2)}{P(2) + P(5) + P(6)} = \frac{.2}{.2 + .1 + .2} = \frac{.2}{.5} = .4$$

$$P(B \mid A) = \frac{P(B \cap A)}{P(A)} = \frac{0}{P(A)} = 0$$

b. Since $P(A \cap B) = 0$, and $P(A) \cdot P(B) > 0$, these two would not be equal, implying A and B are not independent. However, A and B are mutually exclusive, since $P(A \cap B) = 0$.

c. $P(B) = P(2) + P(4) = .2 + .1 = .3$. But $P(B \mid C)$, calculated above, is .4. Since these are not equal, B and C are not independent. Since $P(B \cap C) = .2$, B and C are not mutually exclusive.

3.71 a.

$$\begin{aligned} P(\text{Red}) &= P(\text{Red from Urn 1}) + P(\text{Red from Urn 2}) \\ &= P(\text{Red} \mid \text{Urn 1}) \cdot P(\text{Urn 1}) + P(\text{Red} \mid \text{Urn 2}) \cdot P(\text{Urn 2}) \\ &= \frac{4}{10} \cdot \frac{1}{6} + \frac{6}{10} \cdot \frac{5}{6} = \frac{34}{60} \approx .567 \end{aligned}$$

b.

$$P(\text{Urn 1} \mid \text{Red}) = \frac{P(\text{Urn 1 and Red})}{P(\text{Red})} = \frac{P(\text{Red} \mid \text{Urn 1}) \cdot P(\text{Urn 1})}{P(\text{Red})}$$

$$= \frac{\frac{4}{10} \cdot \frac{1}{6}}{\frac{34}{60}} = \frac{4}{34} \approx .118$$

3.73 Define the following events:

A: {Selected American is highly committed to religion}
B: {Selected American is agnostic or atheist}

a. $P(A) \approx 761/4001 = .1902$

b. $P(B) \approx 299/4001 = .0747$

c. $P(B') \approx 1 - P(B) = 1 - .0747 = .9253$

d. These probabilities were approximated from observed relative frequencies from a sample of only 4001 Americans. We have to assume that the true population relative frequencies are similar to those observed in the sample.

3.75 Define the following events:

A: {Student classified as aggressive}
B: {Student is firstborn}

a. $P(B) = P(B \cap A) + P(B \cap A') = .15 + .25 = .40$

b. $P(A) = P(A \cap B) + P(A \cap B') = .15 + .15 = .30$

c. $P(A \mid B) = \frac{P(A \cap B)}{P(B)} = \frac{.15}{.40} = .375$

d. Since $P(A) \neq P(A \mid B)$, A and B are not independent.

3.77 a. The results of consecutive tosses of a coin are independent. The outcome of one toss probably has no affect on the outcome of the next toss.

b. The opinions of randomly selected individuals are probably independent. The opinion of one individual is not affected by that of another.

c. A major league baseball player's results in two consecutive at-bats are probably dependent. Getting a hit the first time up probably affects what happens the second time up.

d. The amount of gain or loss associated with investments in different stocks bought on the same day and sold on the same day are probably dependent. The stock market has trends. Thus, if one stock is up, the tendency is that another will be up also.

e. The amount of gain or loss associated with investments of different stocks bought and sold at different times are probably independent.

f. The responses of two different individuals to the same stimulus are probably independent. The response of one individual will not affect the response of another.

3.79 a. Let S denote the event that an insect travels toward the pheromone and F the event that an insect travels toward the control. The sample space associated with the experiment of releasing five insects has 32 simple events:

SSSSS	*FSSFS*	*SSFFF*	*FFSFS*
SSSSF	*FSSSF*	*SFSFF*	*FFFSS*
SSSFS	*SFFSS*	*SFFSF*	*FFFFS*
SSFSS	*SFSFS*	*SFFFS*	*FFFSF*
SFSSS	*SFSSF*	*FSSFF*	*FFSFF*
FSSSS	*SSFFS*	*FSFSF*	*FSFFF*
FFSSS	*SSFSF*	*FSFFS*	*SFFFF*
FSFSS	*SSSFF*	*FFSSF*	*FFFFF*

If the pheromone under study has no effect, then each simple event is equally likely and occurs with probability 1/32.

The probability that all five insects travel toward the pheromone is $P(SSSSS) = 1/32 = .03125$.

b. The probability that exactly four of the five insects travel toward the pheromone is:

$$P(SSSSF) + P(SSSFS) + P(SSFSS) + P(SFSSS) + P(FSSSS) = 5/32 = .15625$$

3.81 Define the following events:

C: {Committee's judgment is joint is acceptable}
I: {Inspector's judgment is joint is acceptable}

The simple events of this experiment are:

$C \cap I$
$C \cap I'$
$C' \cap I$
$C' \cap I'$

a. The probability the inspector judges the joint to be acceptable is:

$$P(I) = P(C \cap I) + P(C' \cap I) = \frac{101}{153} + \frac{23}{153} = \frac{124}{153} \approx .810$$

The probability the committee judges the joint to be acceptable is:

$$P(C) = P(C \cap I) + P(C \cap I') = \frac{101}{153} + \frac{10}{153} = \frac{111}{153} \approx .725$$

b. The probability that both the committee and the inspector judge the joint to be acceptable is:

$$P(C \cap I) = \frac{101}{153} \approx .660$$

The probability that neither judge the joint to be acceptable is:

$$P(C' \cap I') = \frac{19}{153} \approx .124$$

c. The probability the inspector and committee disagree is:

$$P(C \cap I') + P(C' \cap I) = \frac{10}{153} + \frac{23}{153} = \frac{33}{153} \approx .216$$

The probability the inspector and committee agree is:

$$P(C \cap I) + P(C' \cap I') = \frac{101}{153} + \frac{19}{153} = \frac{120}{153} \approx .784$$

3.83 Define the following events:

S_1: {Salesman makes sale on the first visit}
S_2: {Salesman makes a sale on the second visit}

$P(S_1) = .4 \qquad P(S_2 | S_1') = .65$

The simple events of the experiment are:

$S_1 \cap S_2'$
$S_1' \cap S_2$
$S_1' \cap S_2'$

The probability the salesman will make a sale is:

$$P(S_1 \cap S_2') + P(S_1' \cap S_2) = P(S_1) + P(S_2 \mid S_1')P(S_1') = .4 + .65(1 - .4) = .4 + .39 = .79$$

3.85 a. The number of ways to draw 2 cards from 52 is:

$$\binom{52}{2} = \frac{52!}{2!(52-2)!} = \frac{52 \cdot 51 \cdot 50 \cdot \cdots \cdot 1}{2 \cdot 1 \cdot 50 \cdot 49 \cdot \cdots \cdot 1} = 1326$$

In a deck of cards, there are 4 aces and 12 face cards. The number of ways to draw 1 ace and 1 face card is:

$$\binom{4}{1}\binom{12}{1} = \frac{4!}{1!(4-1)} \cdot \frac{12!}{1!(12-1)!}$$

$$= \frac{4 \cdot 3 \cdot 2 \cdot 1}{1 \cdot 3 \cdot 2 \cdot 1} \cdot \frac{12 \cdot 11 \cdot 10 \cdot \cdots \cdot 1}{1 \cdot 11 \cdot 10 \cdot 9 \cdot \cdots \cdot 1} = 4(12) = 48$$

Thus, the probability the dealer draws blackjack is $\dfrac{48}{1326} \approx .036$

b. For the player to win with blackjack, the player must draw blackjack while the dealer does not draw blackjack.

The probability the player draws blackjack is $\dfrac{48}{1326}$

Given the player draws blackjack, the number of ways the dealer can draw two cards is:

$$\binom{50}{2} = \frac{50!}{2!(50-2)} = \frac{50 \cdot 49 \cdot 48 \cdot \cdots \cdot 1}{2 \cdot 1 \cdot 48 \cdot 47 \cdot 46 \cdot \cdots \cdot 1} = 1225$$

Given the player draws blackjack, the number of ways the dealer cannot draw blackjack is:

$$\begin{aligned} 1225 - \binom{3}{1}\binom{11}{1} &= 1225 - \frac{3!}{1!(3-1)!} \cdot \frac{11!}{1!(11-1)!} \\ &= 1225 - \frac{3 \cdot 2 \cdot 1}{1 \cdot 2 \cdot 1} \cdot \frac{11 \cdot 10 \cdot 9 \cdot \cdots \cdot 1}{1 \cdot 10 \cdot 9 \cdot 8 \cdot \cdots \cdot 1} \\ &= 1225 - 3(11) = 1192 \end{aligned}$$

Thus, given the player draws blackjack, the probability the dealer will not draw blackjack is:

$$\frac{1192}{1225} \approx .973$$

The probability the player wins with blackjack is the probability the player draws blackjack and the dealer does not, which is:

$$\frac{48}{1326} \cdot \frac{1192}{1225} \approx .035$$

3.87 a. Define the following events:

A_1: {Component 1 works properly}
A_2: {Component 2 works properly}
B_3: {Component 3 works properly}
B_4: {Component 4 works properly}
A: {Subsystem A works properly}
B: {Subsystem B works properly}

The probability a component fails is .1, so the probability a component works properly is $1 - .1 = .9$.

Subsystem A works properly if both components 1 and 2 work properly.

$$P(A) = P(A_1 \cap A_2) = P(A_1)P(A_2) = .9(.9) = .81$$
(since the components operate independently)

Similarly, $P(B) = P(B_1 \cap B_2) = P(B_1)P(B_2) = .9(.9) = .81$

The system operates properly if either subsystem A or B operates properly.

The probability the system operates properly is:

$$\begin{aligned} P(A \cup B) &= P(A) + P(B) - P(A \cap B) = P(A) + P(B) - P(A)P(B) \\ &= .81 + .81 - .81(.81) = .9639 \end{aligned}$$

b. The probability exactly 1 subsystem fails is:

$$\begin{aligned} P(A \cap B') + P(A' \cap B) &= P(A)P(B') + P(A')P(B) \\ &= .81(1 - .81) + (1 - .81).81 = .1539 + .1539 = .3078 \end{aligned}$$

c. The probability the system fails is the probability that both subsystems fail or:

$$P(A' \cap B') = P(A')P(B') = (1 - .81)(1 - .81) = .0361$$

d. The system operates correctly 99% of the time means it fails 1% of the time. The probability 1 subsystem fails is .19. The probability n subsystems fail is $.19^n$. Thus, we must find n such that

$$.19^n \leq .01$$

Thus, $n = 3$.

3.89 Define the events:

A: {A bottle comes from machine A}
B: {A bottle comes from machine B}
R: {A bottle is rejected}.

Then the given probabilities are:

$$P(A) = .75,\ P(B) = .25,\ P(R \mid A) = \frac{1}{20},\ \ P(R \mid B) = \frac{1}{30}$$

The proportion of rejected bottles is:

$$\begin{aligned} P(R) &= P(A \cap R) + P(B \cap R) = P(R|A)P(A) + P(R \mid A)P(B) \\ &= \frac{1}{20}(.75) + \frac{1}{30}(.25) \approx .0458 \end{aligned}$$

The probability that a bottle comes from machine A, given that it is accepted is:

$$P(A \mid R') = \frac{P(A \cap R')}{P(R')} = \frac{P(R' \mid A) \cdot P(A)}{1 - P(R)} = \frac{(19/20) \cdot (.75)}{1 - .0458} \approx .7467$$

Random Variables and Probability Distributions

Chapter 4

4.1 A random variable is a rule that assigns one and only one value to each simple event of an experiment.

4.3 a. Since we can count the number of words spelled correctly, the random variable is discrete.

b. Since we can assume values in an interval, the amount of water flowing through the Hoover Dam in a day is continuous.

c. Since time is measured on an interval, the random variable is continuous.

d. Since we can count the number of bacteria per cubic centimeter in drinking water, this random variable is discrete.

e. Since we cannot count the amount of carbon monoxide produced per gallon of unleaded gas, this random variable is continuous.

f. Since weight is measured on an interval, weight is a continuous random variable.

4.5 a. The reaction time difference is continuous because it lies within an interval.

b. Since we can count the number of violent crimes, this random variable is discrete.

c. Since we can count the number of near misses in a month, this variable is discrete.

d. Since we can count the number of winners each week, this variable is discrete.

e. Since we can count the number of free throws made per game by a basketball team, this variable is discrete.

f. Since distance traveled by a school bus lies in some interval, this is a continuous random variable.

4.7 a. This is a valid distribution because $\sum p(x) = 1$ and $p(x) \geq 0$ for all values of x.

b. This is *not* a valid distribution because $\sum p(x) = .95 \neq 1$.

c. This is *not* a valid distribution because one of the probabilities is negative.

d. The sum of the probabilities over all possible values of the random variable is greater than 1, so this is *not* a valid probability distribution.

4.9 a. The simple events are (where H = head, T = tail):

	HHH	HHT	HTH	THH	HTT	THT	TTH	TTT
x = # heads	3	2	2	2	1	1	1	0

b. If each event is equally likely, then $P(\text{simple event}) = \frac{1}{n} = \frac{1}{8}$

$$p(3) = \frac{1}{8},\ p(2) = \frac{1}{8} + \frac{1}{8} + \frac{1}{8} = \frac{3}{8},\ p(1) = \frac{1}{8} + \frac{1}{8} + \frac{1}{8} = \frac{3}{8},\ \text{and } p(0) = \frac{1}{8}$$

c.

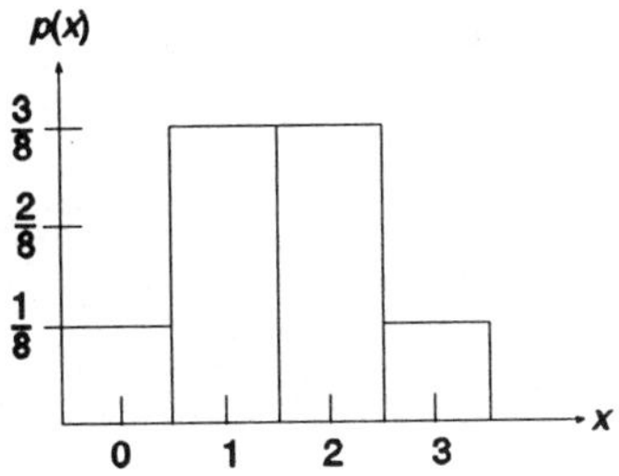

d. $P(x = 2 \text{ or } x = 3) = p(2) + p(3) = \frac{3}{8} + \frac{1}{8} = \frac{4}{8} = \frac{1}{2}$

e. $\mu = E(x) = \sum xp(x) = 0(1/8) + 1(3/8) + 2(3/8) + 3(1/8) = 12/8 = 1.5$
$\sigma^2 = E[(x - \mu)^2] = \sum (x - \mu)^2 p(x)$
$= (0 - 1.5)^2(1/8) + (1 - 1.5)^2(3/8) + (2 - 1.5)^2(3/8) + (3 - 1.5)^2(3/8)$
$= .28125 + .09375 + .09375 + .28125 = .75$

4.11 a. $\mu = E(x) = \sum xp(x) = 1(.2) + 2(.4) + 4(.2) + 10(.2) = .2 + .8 + .8 + 2 = 3.8$

b. $\sigma^2 = E[(x - \mu)^2] = \sum (x - \mu)^2 p(x)$
$= (1 - 3.8)^2(.2) + (2 - 3.8)^2(.4) + (4 - 3.8)^2(.2) + (10 - 3.8)^2(.2)$
$= 1.568 + 1.296 + .008 + 7.688 = 10.56$

c. $\sigma = \sqrt{10.56} = 3.2496$

d. The average value of x over many trials is 3.8.

e. No. The random variable can only take on values 1, 2, 4, or 10.

f. Yes. It is possible that μ can be equal to an actual value of x.

4.13 a. It would seem that the mean of both would be 1 since they both are symmetric distributions centered at 1.

b. $P(x)$ seems more variable since there appears to be greater probability for the two extreme values of 0 and 2 than there is in the distribution of y.

c. For x: $\mu = E(x) = \sum xp(x) = 0(.3) + 1(.4) + 2(.3) = 0 + .4 + .6 = 1$
$\sigma^2 = E[(x - \mu)^2] = \sum (x - \mu)^2 p(x)$
$= (0 - 1)^2(.3) + (1 - 1)^2(.4) + (2 - 1)^2(.3) = .3 + 0 + .3 = .6$

For y: $\mu = E(y) = \sum yp(y) = 0(.1) + 1(.8) + 2(.1) = 0 + .8 + .2 = 1$

$$\sigma^2 = E[(y - \mu)^2] = \sum (y - \mu)^2 p(y)$$
$$= (0 - 1)^2(.1) + (1 - 1)^2(.8) + (2 - 1)^2(.1) = .1 + 0 + .1 = .2$$

The variance for x is larger than that for y.

4.15 a. In order for this to be a valid probability distribution, $\sum p(x) = 1$ and $0 \leq p(x) \leq 1$ for all values of x.

In this distribution, $0 \leq p(x) \leq 1$ for all values of x and $\sum p(x) = .051 + .099 + .093 + .635 + .122 = 1.000$. Thus, this is a valid probability distribution.

b. $P(x = 1) = .051$

c. $P(x \geq 4) = P(x = 4) + P(x = 5) = .635 + .122 = .757$

d. $P(x = 2 \text{ or } x = 3) = P(x = 2) + P(x = 3) = .099 + .093 = .192$

e. $E(x) = \sum xp(x) = 1(.051) + 2(.099) + 3(.093) + 4(.635) + 5(.122) = 3.678$
The average rating for all book reviews is 3.678.

f. $\sigma^2 = E[(x - \mu)^2] = \sum (x - \mu)^2 p(x)$
$= (1 - 3.678)^2(.051) + (2 - 3.678)^2(.099) + (3 - 3.678)^2(.093) + (4 - 3.678)^2(.635) + (5 - 3.678)^2(.122)$
$= .366 + .279 + .043 + .066 + .213 = .967$

$\sigma = \sqrt{\sigma^2} = \sqrt{.967} = .983$

g. The interval is: $\mu \pm 2\sigma \Rightarrow 3.678 \pm 2(.983) \Rightarrow 3.678 \pm 1.966 \Rightarrow (1.712, 5.644)$
$P(1.712 < x < 5.644) = P(x = 2) + P(x = 3) + P(x = 4) + P(x = 5)$
$= .099 + .093 + .635 + .122 = .949$

The Empirical Rule says that approximately .95 of the observations will fall within two standard deviations of the mean. This probability (.949) is very close to this number. Chebyshev's Rule says that at least 3/4 of .75 of the observations will fall within two standard deviations of the mean. Again, .949 is at least .75.

4.17 The sample space of the experiment would be:

S: {*BBB*, *BBG*, *BGB*, *GBB*, *GGB*, *GBG*, *BGG*, *GGG*}

where *B* and *G* represent a boy and girl respectively. Since the female parent always donates an X chromosome, the gender of any child is determined by the chromosome donated by the father; an X chromosome donated by the father will produce a girl, a Y, a boy. Each child therefore has a .5 probability of being either gender (as might be expected). From this, it follows that each of the above simple events has a probability of 1/8. If we let z represent the number of male offspring, the probability distribution of z is:

z	0	1	2	3
$p(z)$	1/8	3/8	3/8	1/8

Then,

$P(\text{At least one boy}) = P(z \geq 1) = 1 - P(z = 0) = 1 - 1/8 = 7/8$

4.19 Let x = bookie's earnings per dollar wagered. Then x can take on values \$1 (you lose) and \$−5 (you win). The only way you win is if you pick 3 winners in 3 games. If the probability of picking 1 winner in 1 game is .5, then $P(www) = p(w)p(w)p(w) = .5(.5)(.5) = .125$ (assuming games are independent).

Thus, the probability distribution for x is:

x	$p(x)$
\$1	.875
\$−5	.125

$E(x) = \sum xp(x) = 1(.875) - 5(.125) = .875 - .625 = \$.25$

4.21 a. The properties of valid probability distributions are:

$$\sum p(x) = 1 \text{ and } 0 \le p(x) \le 1 \text{ for all } x.$$

For Arc a_1: $0 \le p(x) \le 1$ for all x and $\sum p(x) = .6 + .25 + .1 + .05 = 1.00$
Thus, this is a valid probability distribution.

For Arc a_2: $0 \le p(x) \le 1$ for all x and $\sum p(x) = .6 + .3 + .1 = 1.00$
Thus, this is a valid probability distribution.

For Arc a_3: $0 \le p(x) \le 1$ for all x and $\sum p(x) = .9 + .1 = 1.00$
Thus, this is a valid probability distribution.

For Arc a_4: $0 \le p(x) \le 1$ for all x and $\sum p(x) = .9 + .1 = 1.00$
Thus, this is a valid probability distribution.

For Arc a_5: $0 \le p(x) \le 1$ for all x and $\sum p(x) = .9 + .1 = 1.00$
Thus, this is a valid probability distribution.

For Arc a_6: $0 \le p(x) \le 1$ for all x and $\sum p(x) = .7 + .25 + .05 = 1.00$
Thus, this is a valid probability distribution.

b. For Arc a_1, $P(x > 1) = P(x = 2) + P(x = 3) = .25 + .6 = .85$

c. For Arc a_2, $P(x > 1) = P(x = 2) = .6$
For Arc a_3, $P(x > 1) = 0$
For Arc a_4, $P(x > 1) = 0$
For Arc a_5, $P(x > 1) = 0$
For Arc a_6, $P(x > 1) = P(x = 2) = .7$

d. Let x_1 = capacity for Arc a_1 and x_2 = capacity for Arc a_2
$P(x_1 > 1 \text{ and } x_2 > 1) = P(x_1 > 1)P(x_2 > 1) = .85(.6) = .51$

e. For Arc a_1, $E(x) = \sum xp(x) = 3(.6) + 2(.25) + 1(.10) + 0(.05) = 2.4$
The mean capacity for Arc a_1 is 2.4.

For Arc a_2, $E(x) = \sum xp(x) = 2(.60) + 1(.30) + 0(.10) = 1.5$
The mean capacity for Arc a_2 is 1.5.

For Arc a_3, $E(x) = \sum xp(x) = 1(.90) + 0(.10) = 0.9$
The mean capacity for Arc a_3 is 0.9.

For Arc a_4, $E(x) = \sum xp(x) = 1(.90) + 0(.10) = 0.9$
The mean capacity for Arc a_4 is 0.9.

For Arc a_5, $E(x) = \sum xp(x) = 1(.90) + 0(.10) = 0.9$
The mean capacity for Arc a_5 is 0.9.

For Arc a_6, $E(x) = \sum xp(x) = 2(.70) + 1(.25) + 0(.05) = 1.65$
The mean capacity for Arc a_6 is 1.65.

4.23 a. The percent of the sample that indicated they were religious is 22% + 29% + 22% + 19% = 92%. Since this percent is greater than 50%, Newsweek can report that more Americans are religious than not, assuming that the sample is representative of the entire population.

b. $\mu = E(x) = \sum xp(x) = 0(.08) + 1(.22) + 2(.89) + 3(.22) + 4(.19) = 2.22$. The mean religiosity score is 2.22.

4.25 a. $$\frac{6!}{2!(6-2)!} = \frac{6!}{2!4!} = \frac{6 \cdot 5 \cdot 4 \cdot 3 \cdot 2 \cdot 1}{(2 \cdot 1)(4 \cdot 3 \cdot 2 \cdot 1)} = 15$$

b. $$\binom{5}{2} = \frac{5!}{2!(5-2)!} = \frac{5!}{2!3!} = \frac{5 \cdot 4 \cdot 3 \cdot 2 \cdot 1}{(2 \cdot 1)(3 \cdot 2 \cdot 1)} = 10$$

c. $$\binom{7}{0} = \frac{7!}{0!(7-0)!} = \frac{7!}{0!7!} = \frac{7 \cdot 6 \cdot 5 \cdot 4 \cdot 3 \cdot 2 \cdot 1}{(1)(7 \cdot 6 \cdot 5 \cdot 4 \cdot 3 \cdot 2 \cdot 1)} = 1$$
(Note: $0! = 1$)

d. $$\binom{6}{6} = \frac{6!}{6!(6-6)!} = \frac{6!}{6!0!} = \frac{6 \cdot 5 \cdot 4 \cdot 3 \cdot 2 \cdot 1}{(6 \cdot 5 \cdot 4 \cdot 3 \cdot 2 \cdot 1)(1)} = 1$$

e. $$\binom{4}{3} = \frac{4!}{3!(4-3)!} = \frac{4!}{3!1!} = \frac{4 \cdot 3 \cdot 2 \cdot 1}{(3 \cdot 2 \cdot 1)(1)} = 4$$

4.27 a. $$P(x = 1) = \frac{5!}{1!4!}(.2)^1(.8)^4 = \frac{5 \cdot 4 \cdot 3 \cdot 2 \cdot 1}{(1)(4 \cdot 3 \cdot 2 \cdot 1)}(.2)^1(.8)^4 = 5(.2)^1(.8)^4 = .4096$$

b. $$P(x = 2) = \frac{4!}{2!2!}(.6)^2(.4)^2 = \frac{4 \cdot 3 \cdot 2 \cdot 1}{(2 \cdot 1)(2 \cdot 1)}(.6)^2(.4)^2 = 6(.6)^2(.4)^2 = .3456$$

c. $$P(x = 0) = \frac{3!}{0!3!}(.7)^0(.3)^3 = \frac{3 \cdot 2 \cdot 1}{(1)(3 \cdot 2 \cdot 1)}(.7)^0(.3)^3 = 1(.7)^0(.3)^3 = .027$$

d. $P(x = 3) = \frac{5!}{3!2!}(.1)^3(.9)^2 = \frac{5 \cdot 4 \cdot 3 \cdot 2 \cdot 1}{(3 \cdot 2 \cdot 1)(2 \cdot 1)}(.1)^3(.9)^2 = 10(.1)^3(.9)^2 = .0081$

e. $P(x = 2) = \frac{4!}{2!2!}(.4)^2(.6)^2 = \frac{4 \cdot 3 \cdot 2 \cdot 1}{(2 \cdot 1)(2 \cdot 1)}(.4)^2(.6)^2 = 6(.4)^2(.6)^2 = .3456$

f. $P(x = 1) = \frac{3!}{1!2!}(.9)^1(.1)^2 = \frac{3 \cdot 2 \cdot 1}{(1)(2 \cdot 1)}(.9)^1(.1)^2 = 3(.9)^1(.1)^2 = .027$

4.29 a. $\mu = np = 25(.5) = 12.5$

$\sigma^2 = np(1 - p) = 25(.5)(.5) = 6.25$

$\sigma = \sqrt{\sigma^2} = \sqrt{6.25} = 2.5$

b. $\mu = np = 80(.2) = 16$

$\sigma^2 = np(1 - p) = 80(.2)(.8) = 12.8$

$\sigma = \sqrt{\sigma^2} = \sqrt{12.8} = 3.578$

c. $\mu = np = 100(.6) = 60$

$\sigma^2 = np(1 - p) = 100(.6)(.4) = 24$

$\sigma = \sqrt{\sigma^2} = \sqrt{24} = 4.899$

d. $\mu = np = 70(.9) = 63$

$\sigma^2 = np(1 - p) = 70(.9)(.1) = 6.3$

$\sigma = \sqrt{\sigma^2} = \sqrt{6.3} = 2.510$

e. $\mu = np = 60(.8) = 48$

$\sigma^2 = np(1 - p) = 60(.8)(.2) = 9.6$

$\sigma = \sqrt{\sigma^2} = \sqrt{9.6} = 3.098$

f. $\mu = np = 1{,}000(.04) = 40$

$\sigma^2 = np(1 - p) = 1{,}000(.04)(.96) = 38.4$

$\sigma = \sqrt{\sigma^2} = \sqrt{38.4} = 6.197$

4.31 a. The simple events listed below are all equally likely, implying a probability of 1/32 for each. The list is in a regular pattern such that the first simple event would yield $x = 0$, the next five yield $x = 1$, the next ten yield $x = 2$, the next ten also yield $x = 3$, the next five yield $x = 4$, and the final one yields $x = 5$. The resulting probability distribution is given below the simple events.

FFFFF, FFFFS, FFFSF, FFSFF, FSFFF, SFFFF, FFFSS, FFSFS
FSFFS, SFFFS, FFSSF, FSFSF, SFFSF, FSSFF, SFSFF, SSFFF
FFSSS, FSFSS, SFFSS, FSSFS, SFSFS, SSFFS, FSSSF, SFSSF
SSFSF, SSSFF, FSSSS, SFSSS, SSFSS, SSSFS, SSSSF, SSSSS

x	0	1	2	3	4	5
$p(x)$	1/32	5/32	10/32	10/32	5/32	1/32

b. $$P(x = 0) = \frac{5!}{0!5!}\left(\frac{1}{2}\right)^0\left(\frac{1}{2}\right)^5 = \frac{5 \cdot 4 \cdot 3 \cdot 2 \cdot 1}{(1)(5 \cdot 4 \cdot 3 \cdot 2 \cdot 1)}\left(\frac{1}{2}\right)^0\left(\frac{1}{2}\right)^5 = 1\left(\frac{1}{2}\right)^0\left(\frac{1}{2}\right)^5 = \frac{1}{32} = .03125$$

$$P(x = 1) = \frac{5!}{1!4!}\left(\frac{1}{2}\right)^1\left(\frac{1}{2}\right)^4 = \frac{5 \cdot 4 \cdot 3 \cdot 2 \cdot 1}{(1)(4 \cdot 3 \cdot 2 \cdot 1)}\left(\frac{1}{2}\right)^1\left(\frac{1}{2}\right)^4 = 5\left(\frac{1}{2}\right)^1\left(\frac{1}{2}\right)^4 = \frac{5}{32} = .15625$$

$$P(x = 2) = \frac{5!}{2!3!}\left(\frac{1}{2}\right)^2\left(\frac{1}{2}\right)^3 = \frac{5 \cdot 4 \cdot 3 \cdot 2 \cdot 1}{(2 \cdot 1)(3 \cdot 2 \cdot 1)}\left(\frac{1}{2}\right)^2\left(\frac{1}{2}\right)^3 = 10\left(\frac{1}{2}\right)^2\left(\frac{1}{2}\right)^3 = \frac{10}{32} = .3125$$

$$P(x = 3) = \frac{5!}{3!2!}\left(\frac{1}{2}\right)^3\left(\frac{1}{2}\right)^2 = \frac{5 \cdot 4 \cdot 3 \cdot 2 \cdot 1}{(3 \cdot 2 \cdot 1)(2 \cdot 1)}\left(\frac{1}{2}\right)^3\left(\frac{1}{2}\right)^2 = 10\left(\frac{1}{2}\right)^3\left(\frac{1}{2}\right)^2 = \frac{10}{32} = .3125$$

$$P(x = 4) = \frac{5!}{4!1!}\left(\frac{1}{2}\right)^4\left(\frac{1}{2}\right)^1 = \frac{5 \cdot 4 \cdot 3 \cdot 2 \cdot 1}{(4 \cdot 3 \cdot 2 \cdot 1)(1)}\left(\frac{1}{2}\right)^4\left(\frac{1}{2}\right)^1 = 5\left(\frac{1}{2}\right)^4\left(\frac{1}{2}\right)^1 = \frac{5}{32} = .15625$$

$$P(x = 5) = \frac{5!}{5!0!}\left(\frac{1}{2}\right)^5\left(\frac{1}{2}\right)^0 = \frac{5 \cdot 4 \cdot 3 \cdot 2 \cdot 1}{(5 \cdot 4 \cdot 3 \cdot 2 \cdot 1)(1)}\left(\frac{1}{2}\right)^5\left(\frac{1}{2}\right)^0 = 1\left(\frac{1}{2}\right)^5\left(\frac{1}{2}\right)^0 = \frac{1}{32} = .03125$$

c. From Table II, $n = 5, p = .5$:

$P(x = 0) = .031$
$P(x = 1) = .188 - .031 = .157$
$P(x = 2) = .500 - .188 = .312$
$P(x = 3) = .812 - .500 = .312$
$P(x = 4) = .969 - .812 = .157$
$P(x = 5) = 1 - .969 = .031$

4.33 In order for x to have a binomial distribution, it must possess the following characteristics:

1. *n* identical trials. A sample of 608 married women is fairly small compared to the entire population of married women, so the 608 trials would be very close to identical.
2. There are only two possible outcomes for each trial. In this experiment, there are only two possible outcomes: the woman would marry the same man (*S*) or she would not (*F*).

3. The probability of Success stays the same from trial to trial. Here, $P(S) = .80$ and $P(F) = .20$. In reality, these probabilities would not be exactly the same from trial to trial, but rounded off to 4 decimal places, they would be the same.
4. The trials are independent. In this experiment, the trials would not be exactly independent because we would be sampling without replacement from a finite population. However, if the sample of size 608 is fairly small compared to the population size, the trials will be essentially independent.
5. The binomial random variable x would be the number of successes in n trials. For this experiment, $x =$ the number of women out of the 608 who would marry the same man again.

Thus, x would possess (approximately) a binomial distribution.

4.35 a. Yes, x can be treated as a binomial.

1. There are $n = 36$ trials.
2. There are only two outcomes for each trial: $S =$ golfer makes a hole-in-one and $F =$ golfer does not make a hole-in-one.
3. The probability of success is the same for each golfer: $p = 1/2780$; $q = 2779/2780$
4. We must assume that the trials are independent—i.e., that the outcome of one golfer is not related to the outcome of any other golfer.
5. $x =$ number of golfers making a hole-in-one in 36 trials.

b. $$P(x = 0) = \binom{36}{0}(1/2780)^0(2779/2780)^{36} = .9873$$

c. $$P(x = 4) = \binom{36}{4}(1/2780)^4(2779/2780)^{32} = 9.749 \times 10^{-10}$$

4.37 a. For this problem, we must assume that each couple has only one child. Let $x =$ number of couples out of 15 who have a child with A-T. Then x is a binomial random variable with $n = 15$, $S =$ couple has child with A-T, $F =$ couple does not have a child with A-T, $P(S) = p = 1/4 = .25$ and $q = 1 - p = 1 - .25 = .75$.

$$\begin{aligned}P(x > 10) &= P(x = 11) + P(x = 12) + P(x = 13) + P(x = 14) + P(x = 15)\\ &= \frac{15!}{11!4!}.25^{11}.75^4 + \frac{15!}{12!3!}.25^{12}.75^3 + \frac{15}{13!2!}.25^{13}.75^2\\ &\quad + \frac{15!}{14!1!}.25^{14}.75^1 + \frac{15!}{15!0!}.25^{15}.75^0\\ &= .00010 + .00001 + .00000 + .00000 + .00000 = .00011\end{aligned}$$

b. For $n = 10{,}000$,

$$\begin{aligned}P(x < 3{,}000) &= P(x = 0) + P(x = 1) + \cdots + P(x = 2{,}999)\\ &= \frac{10000!}{0!10000!}.25^0.75^{10000} + \frac{10000!}{1!9999!}.25^1.75^{9999} + \cdots + \frac{10000!}{2999!7001!}.25^{2999}.75^{7001}\end{aligned}$$

$\mu = E(x) = np = 10{,}000(.25) = 2{,}500$; $\sigma^2 = npq = 10{,}000(.25)(.75) = 1{,}875$

$\sigma = \sqrt{1{,}875} = 43.301$

The value 3,000 has a z-score of:

$$z = \frac{x - \mu}{\sigma} = \frac{3{,}000 - 2{,}500}{43.301} = 11.547$$

This z-score is extremely large. Essentially all of the observations will have z-scores less than 11.547. Thus, $P(x < 3000) \approx 1.000$.

4.39 Define x as the number of components that operate successfully. The random variable x is a binomial random variable (the components operate independently and there are only two possible outcomes) with $n = 4$ and $p = .85$.

$$P(\text{system fails}) = P(x = 0) = \binom{4}{0} .85^0(.15)^{4-0} = \frac{4!}{0!4!} .85^0(.15)^4 = .15^4 = .0005$$

4.41 a. Since there are 5 different cards, the probability of guessing correctly is $1/5 = .20$ since they all have an equal chance of occurring.

b. The random variable x is a binomial random variable with $n = 20$ and $p = .20$ (since the psychic is guessing).

$$\mu = E(x) = 20(.20) = 4$$

c. $$\begin{aligned} P(x \geq 6) &= 1 - P(x \leq 5) \\ &= 1 - .804 \quad \text{(Table II, Appendix A)} \\ &= .196 \end{aligned}$$

d. If the psychic is guessing, the probability of getting at least 6 correct is .196. Since there is a substantial probability that a person can get six or more correct by guessing, this evidence does not indicate that the psychic is doing anything more than guessing.

4.43 a. $\mu = E(x) = np = 800(.65) = 520$

$\sigma = \sqrt{npq} = \sqrt{800(.65)(.35)} = 13.491$

b. Half of the 800 food items is 400. A value of $x = 400$ would have a z-score of:

$$z = \frac{x - \mu}{\sigma} = \frac{400 - 520}{13.491} = -8.895$$

Since the z-score associated with 400 items is so small (-8.895), it would be virtually impossible to observe less than half without any traces of pesticides if the 65% value was correct.

4.45 a. $$\begin{aligned} P(-1 \leq z \leq 1) &= A_1 + A_2 \\ &= .3413 + .3413 \\ &= .6826 \end{aligned}$$

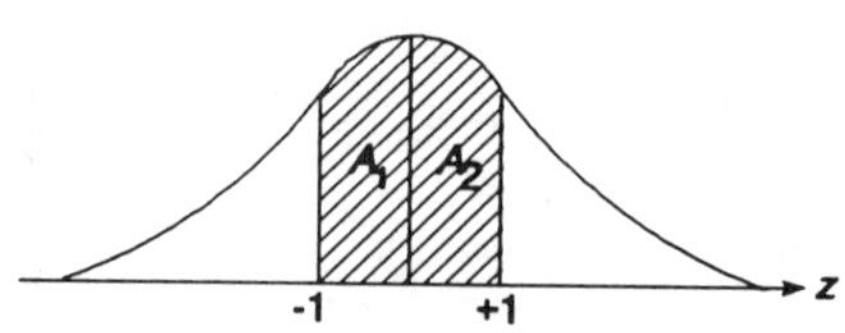

b. $P(-2 \le z \le 2) = A_1 + A_2$
$= .4772 + .4772$
$= .9544$

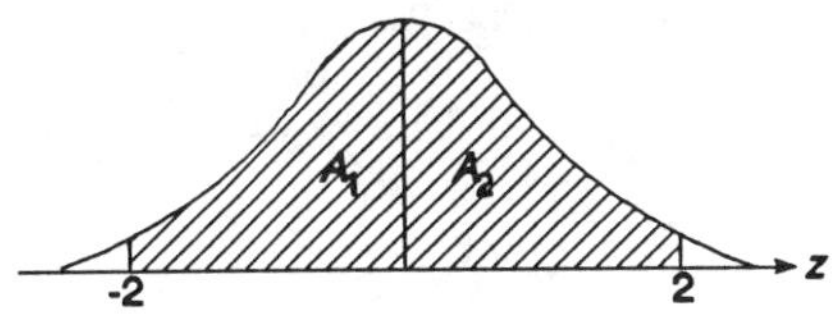

c. $P(-2.16 < z \le 0.55) = A_1 + A_2$
$= .4846 + .2088$
$= .6934$

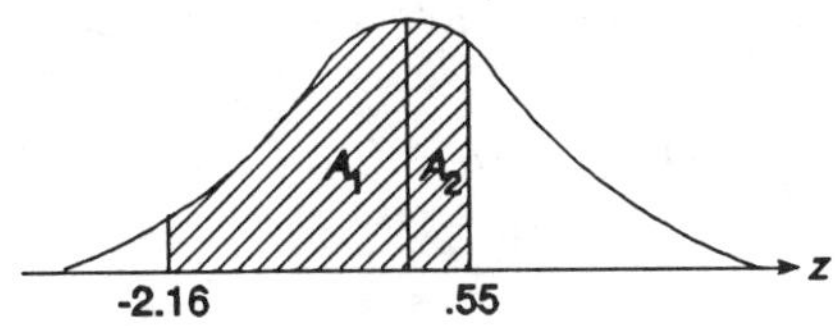

d. $P(-.42 < z < 1.96)$
$= P(-.42 \le z \le 0) + P(0 \le z \le 1.96)$
$= A_1 + A_2$
$= .1628 + .4750$
$= .6378$

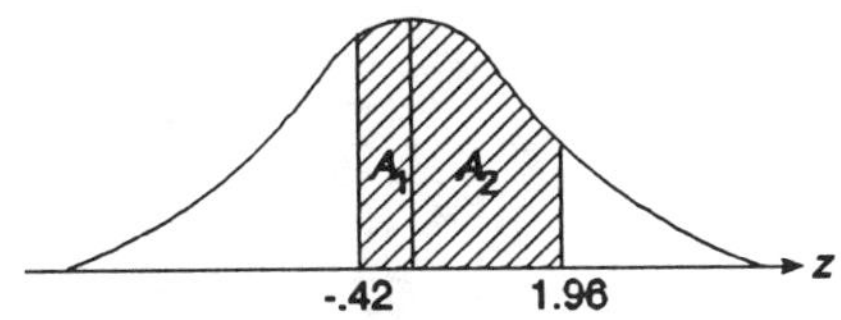

e. $P(z \ge -2.33) = P(-2.33 \le z \le 0) + P(z \ge 0)$
$= A_1 + A_2$
$= .4901 + .5000$
$= .9901$

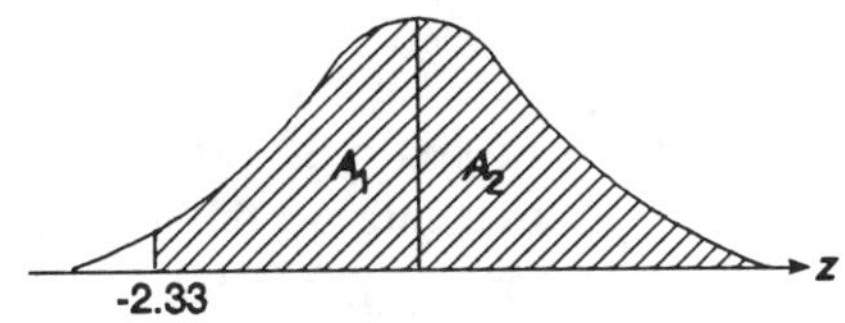

f. $P(z < 2.33) = P(z \le 0) + P(0 \le z \le 2.33)$
$= A_1 + A_2$
$= .5000 + .4901$
$= .9901$

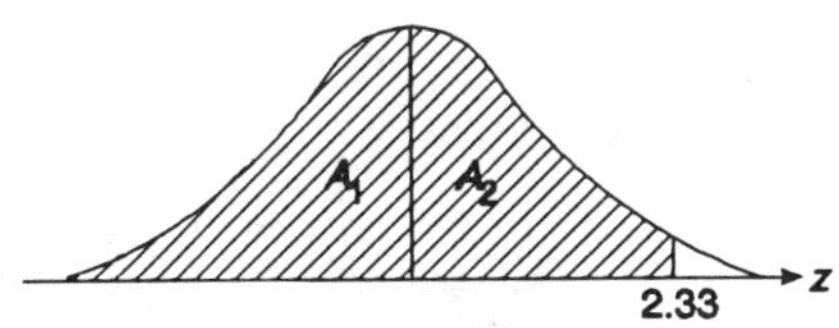

4.47 Using Table III, Appendix A:

a. $P(-1 \le z \le 1)$
$= P(-1 \le z \le 0) + P(0 < z \le 1)$
$= .3413 + .3413 = .6826$

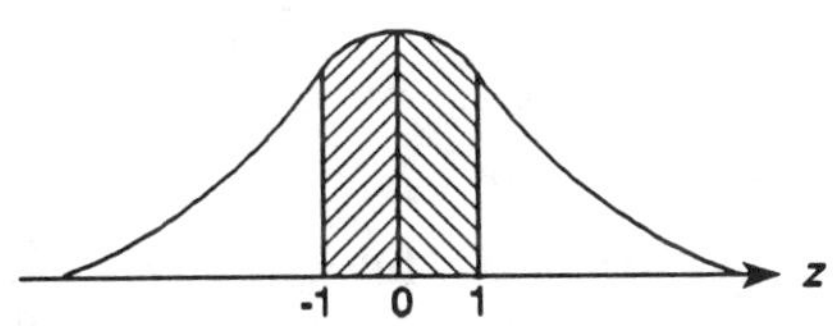

b. $P(-1.96 \le z \le 1.96)$
$= P(-1.96 \le z < 0) + P(0 \le z \le 1.96)$
$= .4750 + .4750 = .9500$

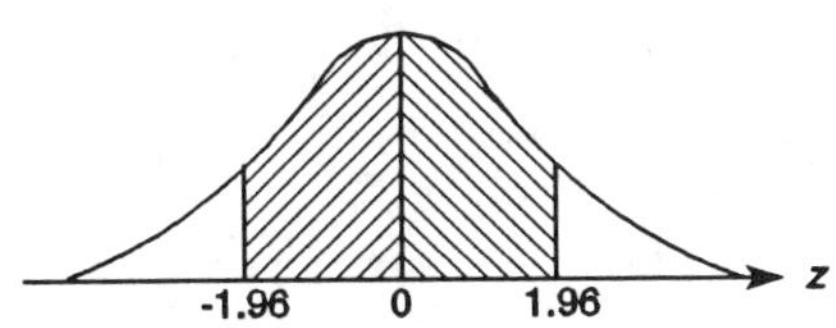

c. $P(-1.645 \le z \le 1.645)$
$= P(-1.645 \le z < 0) + P(0 \le z \le 1.645)$
$= .4500 + .4500 = .90$
(using interpolation)

d. $P(-2 \le z \le 2)$
$= P(-2 \le z < 0) + P(0 \le z \le 2)$
$= .4772 + .4772 = .9544$

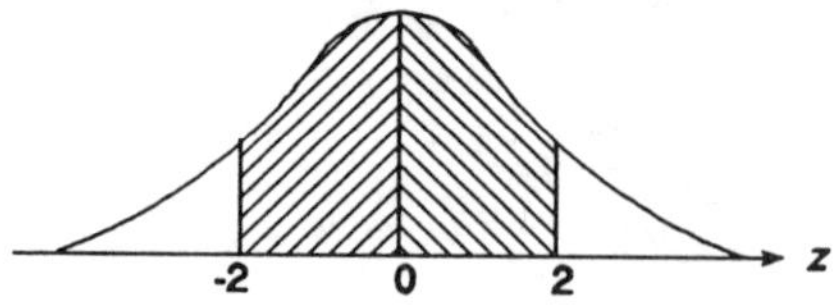

4.49 Using Table III, Appendix A:

a. $P(z \ge z_0) = .5$
$z_0 = 0$

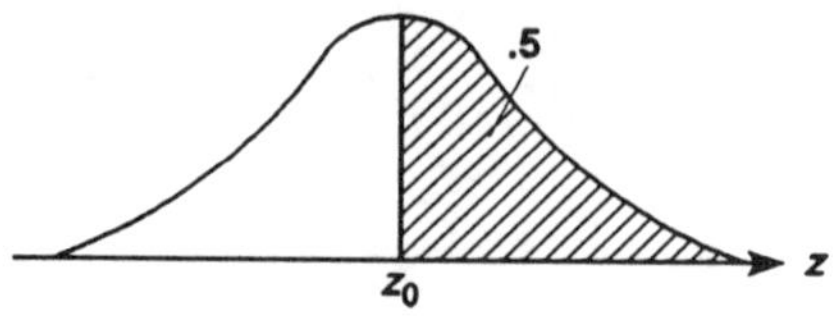

b. $P(z \ge z_0) = .0057$
$A_1 = .5 - .0057 = .4943$
Looking up the area .4943 in Table III gives $z_0 = 2.53$.

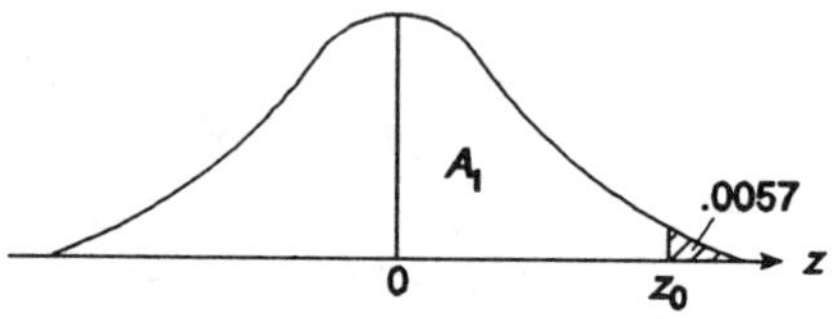

c. $P(0 \le z \le z_0) = .4713$
Looking up the area .4713 in Table III gives $z_0 = 1.90$.

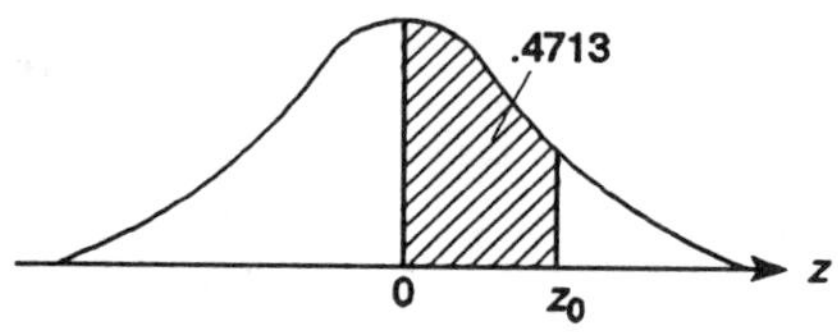

d. $P(z < z_0) = .0392$
$A_1 = .5 - .0392 = .4608$
Looking up area .4608 in Table III gives $z = 1.76$. Because z_0 is to the left of 0, $z_0 = -1.76$.

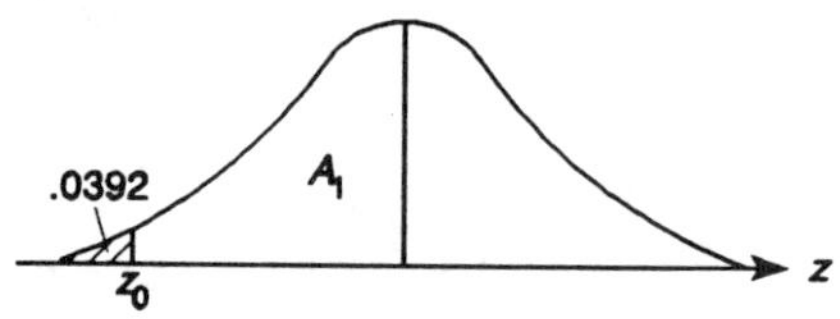

4.51 Using the formula $z = \dfrac{x - \mu}{\sigma}$ with $\mu = 25$ and $\sigma = 5$:

a. $z = \dfrac{25 - 25}{5} = 0$

b. $z = \dfrac{30 - 25}{5} = 1$

c. $z = \dfrac{37.5 - 25}{5} = 2.5$

d. $z = \dfrac{10 - 25}{5} = -3$

e. $z = \dfrac{50 - 25}{5} = 5$

f. $z = \frac{32 - 25}{5} = 1.4$

4.53 Using Table III, Appendix A:

a. $P(x \le 50) = P\left[z \le \frac{50 - 40}{10}\right] = P(z \le 1) = .5 + .3413 = .8413$

b. $P(x \le 35.6) = P\left[z \le \frac{35.6 - 40}{10}\right] = P(z \le -.44) = .5 - .1700 = .3300$

c. $P(35 \le x \le 56.8) = P\left[\frac{35 - 40}{10} \le z \le \frac{56.8 - 40}{10}\right]$
$= P(-.5 \le z \le 1.68) = .1915 + .4535 = .6450$

d. $P(22.9 \le x \le 33.2) = P\left[\frac{22.9 - 40}{10} \le z \le \frac{33.2 - 40}{10}\right]$
$= P(-1.71 \le z \le -.68) = .4564 - .2517 = .2047$

e. $P(x \ge 25.3) = P\left[z \ge \frac{25.3 - 40}{10}\right] = P(z \ge -1.47) = .5 + .4292 = .9292$

f. $P(x \le 25.3) = P(z \le -1.47) = .5 - .4292 = .0708$

4.55

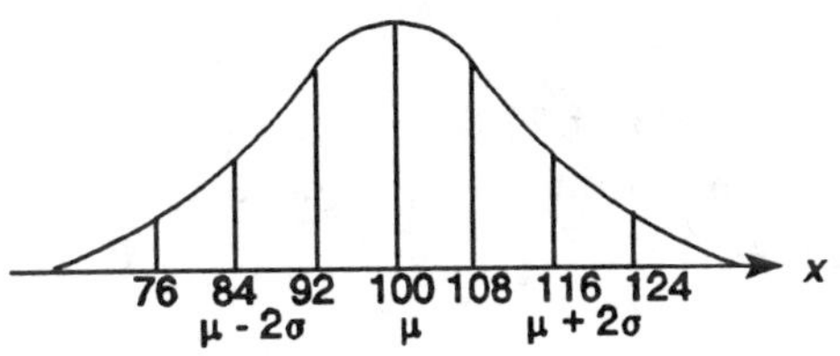

Using Table III, Appendix A:

a. $P(\mu - 2\sigma \le x \le \mu + 2\sigma) = P(-2 \le z \le 2)$
$= P(-2 \le z \le 0) + P(0 \le z \le 2)$
$= .4772 + .4772 = .9544$

b. $P(x \ge \mu + 2\sigma) = P(z \ge 2) = .5 - .4772 = .0228$

c. $P(x \le 92) = P\left[z \le \frac{92 - 100}{8}\right] = P(z \le -1) = .5 - .3413 = .1587$

d. $P(92 \le x \le 116) = P\left[\frac{92 - 100}{8} \le z \le \frac{116 - 100}{8}\right] = P(-1 \le z \le 2)$
$= P(-1 \le z \le 0) + P(0 \le z \le 2)$
$= .3413 + .4772 = .8185$

e. $P(92 \le x \le 96) = P\left[\frac{92 - 100}{8} \le z \le \frac{96 - 100}{8}\right] = P(-1 \le z \le -.5)$

$= P(-1 \le z \le 0) - P(-.5 \le z \le 0)$
$= .3413 - .1915 = .1498$

f. $P(76 \le x \le 124) = P\left[\frac{76 - 100}{8} \le z \le \frac{124 - 100}{8}\right] = P(-3 \le z \le 3)$

$= P(-3 \le z \le 0) + P(0 \le z \le 3)$
$= .4987 + .4987 = .9974$

4.57 a. Using Table III, Appendix A, with $\mu = 20.2$ and $\sigma = .65$,

$$P(20 < x < 21) = P\left[\frac{20 - 20.2}{.65} < z < \frac{21 - 20.2}{.65}\right] = P(-.31 < z < 1.23)$$
$$= P(-.31 < z < 0) + P(0 \le z < 1.23) = .1217 + .3907 = .5124$$

b. $P(x \le 19.84) = P\left[z \le \frac{19.84 - 20.2}{.65}\right] = P(z \le -.55) = .5 - .2088 = .2912$

Since the probability of observing a sardine with a length of 19.84 cm or smaller is not small ($p = .2912$), this is not an unusual event if the sardine was, in fact, two years old. Thus, it is likely that the sardine is two years old.

c. $P(x \ge 22.01) = P\left[z \ge \frac{22.01 - 20.2}{.65}\right] = P(z \ge 2.78) = .5 - .4973 = .0027$

Since the probability of observing a sardine with a length of 22.01 cm or larger is so small ($p = .0027$), this would be a very unusual event if the sardine was, in fact, two years old. Thus, it is unlikely that the sardine is two years old.

4.59 a. Using Table III, Appendix A, and $\mu = 75$ and $\sigma = 7.5$,

$$P(x > 80) = P\left[z > \frac{80 - 75}{7.5}\right] = P(z > .67) = .5 - .2486 = .2514$$

Thus, 25.14% of the scores exceeded 80.

b. $P(x \le x_0) = .98$. Find x_0.

$P(x \le x_0) = P\left[z \le \frac{x_0 - 75}{7.5}\right] = P(z \le z_0) = .98$

$A_1 = .98 - .5 = .4800$
Looking up area .4800 in Table III, $z_0 = 2.05$.

$z_0 = \frac{x_0 - 75}{7.5} \Rightarrow 2.05 = \frac{x_0 - 75}{7.5} \Rightarrow x_0 = 90.375$

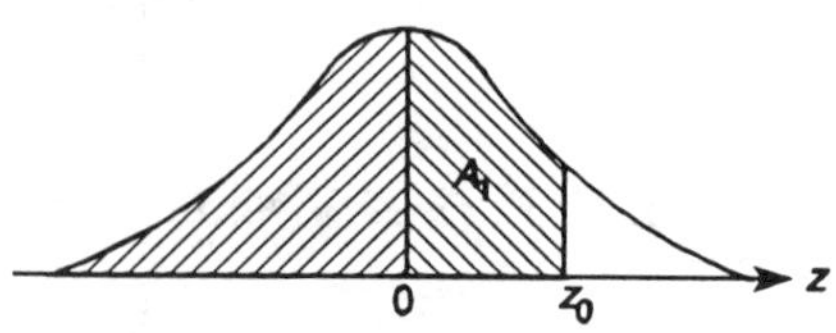

4.61 a. Using Table III, Appendix A, with $\mu = 8.72$ and $\sigma = 1.10$,

$$P(x < 6) = P\left(z < \frac{6 - 8.72}{1.10}\right) = P(z < -2.47) = .5 - .4932 = .0068$$

Thus, approximately .68% of the games would result in fewer than 6 hits.

b. The probability of observing fewer than 6 hits in a game is $p = .0068$. The probability of observing 0 hits would be even smaller. Thus, it would be extremely unusual to observe a no hitter.

4.63 From Table III, Appendix A, with $\mu = 6.3$ and $\sigma = .6$

a. $$P(x < 5) = P\left(z < \frac{5 - 6.3}{.6}\right) = P(z < -2.17) = .5 - .4850 = .0150$$

Thus, the percentage of days when the oxygen content is undesirable is 1.5%.

b. We would expect the oxygen contents to fall within 2 standard deviations of the mean, or

$$\mu \pm 2\sigma \Rightarrow 6.3 \pm 2(.6) \Rightarrow 6.3 \pm 1.2 \Rightarrow (5.1, 7.5)$$

4.65 a. Using Table III, Appendix A, with $\mu = 450$ and $\sigma = 40$,

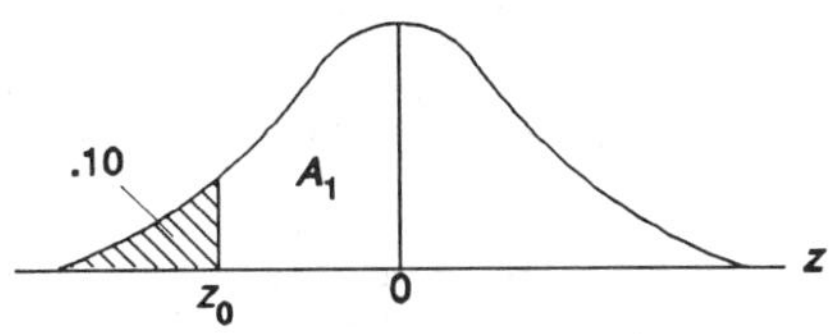

$$P(x < x_0) = .10 \Rightarrow P\left(z < \frac{x_0 - 450}{40}\right)$$
$$= P(z < z_0) = .10$$

$A_1 = .5 - .10 = .4000$
Looking up the area .4000 in Table III gives $z_0 = 1.28$. Since z_0 is to the left of 0, $z_0 = -1.28$.

$$z_0 = -1.28 = \frac{x_0 - 450}{40} \Rightarrow x_0 = 40(-1.28) + 450 = 398.8 \text{ seconds.}$$

4.67 a. Using Table III, Appendix A, with $\mu = 99$ and $\sigma = 4.3$,

$$P(x > x_0) = .99 \Rightarrow P\left(z < \frac{x_0 - 99}{4.3}\right)$$
$$= P(z > z_0) = .99$$

$A_1 = .99 - .5 = .4900$
Looking up the area .4900 in Table III gives $z_0 = 2.33$. Since z_0 is to the left of 0, $z_0 = -2.33$.

$$z_0 = -2.33 = \frac{x_0 - 99}{4.3} \Rightarrow x_0 = 4.3(-2.33) + 99 = 88.981$$

4.69 a–b. The different samples of $n = 2$ with replacement and their means are:

Possible Samples	$\bar{x}$	Possible Samples	$\bar{x}$
0, 0	0	4, 0	2
0, 2	1	4, 2	3
0, 4	2	4, 4	4
0, 6	3	4, 6	5
2, 0	1	6, 0	3
2, 2	2	6, 2	4
2, 4	3	6, 4	5
2, 6	4	6, 6	6

c. Since each sample is equally likely, the probability of any 1 being selected is $\frac{1}{4}\left(\frac{1}{4}\right) = \frac{1}{16}$

d. $P(\bar{x} = 0) = \frac{1}{16}$

$P(\bar{x} = 1) = \frac{1}{16} + \frac{1}{16} = \frac{2}{16}$

$P(\bar{x} = 2) = \frac{1}{16} + \frac{1}{16} + \frac{1}{16} = \frac{3}{16}$

$P(\bar{x} = 3) = \frac{1}{16} + \frac{1}{16} + \frac{1}{16} + \frac{1}{16} = \frac{4}{16}$

$P(\bar{x} = 4) = \frac{1}{16} + \frac{1}{16} + \frac{1}{16} = \frac{3}{16}$

$P(\bar{x} = 5) = \frac{1}{16} + \frac{1}{16} = \frac{2}{16}$

$P(\bar{x} = 6) = \frac{1}{16}$

$\bar{x}$	$p(\bar{x})$
0	1/16
1	2/16
2	3/16
3	4/16
4	3/16
5	2/16
6	1/16

e.

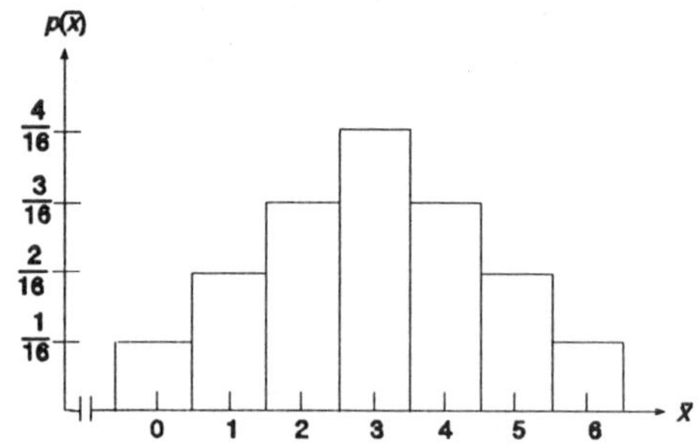

4.71 If the observations are independent of each other, then

$P(1, 1) = p(1)p(1) = .2(.2) = .04$
$P(1, 2) = p(1)p(2) = .2(.3) = .06$
$P(1, 3) = p(1)p(3) = .2(.2) = .04$
etc.

a.

Possible Samples	$\bar{x}$	$p(\bar{x})$	Possible Samples	$\bar{x}$	$p(\bar{x})$
1, 1	1	.04	3, 4	3.5	.04
1, 2	1.5	.06	3, 5	4	.02
1, 3	2	.04	4, 1	2.5	.04
1, 4	2.5	.04	4, 2	3	.06
1, 5	3	.02	4, 3	3.5	.04
2, 1	1.5	.06	4, 4	4	.04
2, 2	2	.09	4, 5	4.5	.02
2, 3	2.5	.06	5, 1	3	.02
2, 4	3	.06	5, 2	3.5	.03
2, 5	3.5	.03	5, 3	4	.02
3, 1	2	.04	5, 4	4.5	.02
3, 2	2.5	.06	5, 5	5	.01
3, 3	3	.04			

Summing the probabilities, the probability distribution of $\bar{x}$ is:

$\bar{x}$	$p(\bar{x})$
1	.04
1.5	.12
2	.17
2.5	.20
3	.20
3.5	.14
4	.08
4.5	.04
5	.01

b.

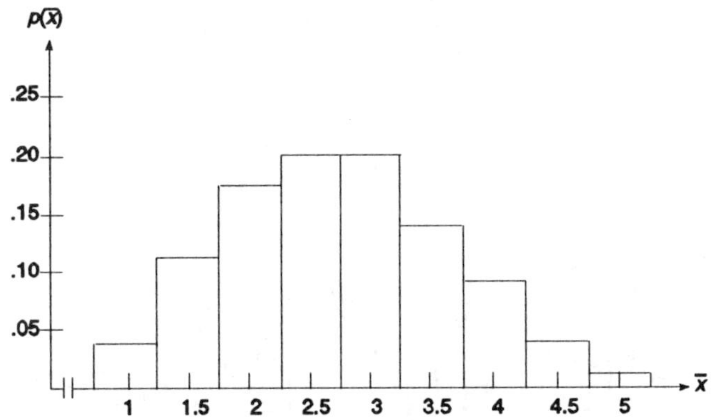

c. $P(\bar{x} \geq 4.5) = .04 + .01 = .05$

d. No. The probability of observing $\bar{x} = 4.5$ or larger is small (.05).

4.73 a. For a sample of size $n = 2$, the sample mean and sample median are exactly the same. Thus, the sampling distribution of the sample median is the same as that for the sample mean (see Exercise 4.71a).

b. The probability histogram for the sample median is identical to that for the sample mean (see Exercise 4.71b).

4.77 a. $\mu = \sum xp(x) = 2\left(\frac{1}{3}\right) + 4\left(\frac{1}{3}\right) + 9\left(\frac{1}{3}\right) = \frac{15}{3} = 5$

b. The possible samples of size $n = 3$, the sample means, and the probabilities are:

Possible Samples	$\bar{x}$	$p(\bar{x})$	m	Possible Samples	$\bar{x}$	$p(\bar{x})$	m
2, 2, 2	2	1/27	2	4, 4, 4	4	1/27	4
2, 2, 4	8/3	1/27	2	4, 4, 9	17/3	1/27	4
2, 2, 9	13/3	1/27	2	4, 9, 2	5	1/27	4
2, 4, 2	8/3	1/27	2	4, 9, 4	17/3	1/27	4
2, 4, 4	10/3	1/27	4	4, 9, 9	22/3	1/27	9
2, 4, 9	5	1/27	4	9, 2, 2	13/3	1/27	2
2, 9, 2	13/3	1/27	2	9, 2, 4	5	1/27	4
2, 9, 4	5	1/27	4	9, 2, 9	20/3	1/27	9
2, 9, 9	20/3	1/27	9	9, 4, 2	5	1/27	4
4, 2, 2	8/3	1/27	2	9, 4, 4	17/3	1/27	4
4, 2, 4	10/3	1/27	4	9, 4, 9	22/3	1/27	9
4, 2, 9	5	1/27	4	9, 9, 2	20/3	1/27	9
4, 4, 2	10/3	1/27	4	9, 9, 4	22/3	1/27	9
				9, 9, 9	9	1/27	9

The sampling distribution of $\bar{x}$ is:

$\bar{x}$	$p(\bar{x})$
2	1/27
8/3	3/27
10/3	3/27
4	1/27
13/3	3/27
5	6/27
17/3	3/27
20/3	3/27
22/3	3/27
9	1/27
	27/27

$$E(\bar{x}) = \sum \bar{x}p(\bar{x}) = 2\left(\frac{1}{27}\right) + \frac{8}{3}\left(\frac{3}{27}\right) + \frac{10}{3}\left(\frac{3}{27}\right) + 4\left(\frac{1}{27}\right) + \frac{13}{3}\left(\frac{3}{27}\right)$$

$$+ 5\left(\frac{6}{27}\right) + \frac{17}{3}\left(\frac{3}{27}\right) + \frac{20}{3}\left(\frac{3}{27}\right) + \frac{22}{3}\left(\frac{3}{27}\right) + 9\left(\frac{1}{27}\right)$$

$$= \frac{2}{27} + \frac{8}{27} + \frac{10}{27} + \frac{4}{27} + \frac{13}{27} + \frac{30}{27} + \frac{17}{27} + \frac{20}{27} + \frac{22}{27} + \frac{9}{27}$$

$$= \frac{135}{27} = 5$$

Since $\mu = 5$ in part **a**, and $E(\bar{x}) = \mu = 5$, $\bar{x}$ is an unbiased estimator of μ.

c. The median was calculated for each sample and is shown in the table in part **b**. The sampling distribution of m is:

m	$p(m)$
2	7/27
4	13/27
9	7/27
	27/27

$$E(m) = \sum mp(m) = 2\left(\frac{7}{27}\right) + 4\left(\frac{13}{27}\right) + 9\left(\frac{7}{27}\right) = \frac{14}{27} + \frac{52}{27} + \frac{63}{27} = \frac{129}{27} = 4.778$$

The $E(m) = 4.778 \neq \mu = 5$. Thus, m is a biased estimator of μ.

d. Use the sample mean, $\bar{x}$. It is an unbiased estimator.

4.81 a. Refer to the solution to Exercise 4.71. The values of s^2 and the corresponding probabilities are listed below:

$$s^2 = \frac{\sum (x^2) - \frac{(\sum x)^2}{n}}{n - 1}$$

For sample 1, 1, $s^2 = \frac{2 - \frac{2^2}{2}}{1} = 0$

For sample 1, 2, $s^2 = \frac{5 - \frac{3^2}{2}}{1} = .5$

The rest of the values are calculated and shown:

s^2	$p(s^2)$	s^2	$p(s^2)$
0.0	.04	0.5	.04
0.5	.06	2.0	.02
2.0	.04	4.5	.04
4.5	.04	2.0	.06
8.0	.02	0.5	.04
0.5	.06	0.0	.04
0.0	.09	0.5	.02
0.5	.06	8.0	.02
2.0	.06	4.5	.03
4.5	.03	2.0	.02
2.0	.04	0.5	.02
0.5	.06	0.0	.01
0.0	.04		

The sampling distribution of s^2 is:

s^2	$p(s^2)$
0.0	.22
0.5	.36
2.0	.24
4.5	.14
8.0	.04

b. $\sigma^2 = \sum (x - \mu)^2 p(x) = (1 - 2.7)^2(.2) + (2 - 2.7)^2(.3) + (3 - 2.7)^2(.2) + (4 - 2.7)^2(.2) + (5 - 2.7)^2(.1) = 1.61$

c. $E(s^2) = \sum s^2 p(s^2) = 0(.22) + .5(.36) + 2(.24) + 4.5(.14) + 8(.04) = 1.61$

d. The sampling distribution of s is listed below, where $s = \sqrt{s^2}$:

s	$p(s)$
0.000	.22
0.707	.36
1.414	.24
2.121	.14
2.828	.04

e. $E(s) = \sum sp(s) = 0(.22) + .707(.36) + 1.414(.24) + 2.121(.14) + 2.828(.04) = 1.00394$

Since $E(s) = 1.00394$ is not equal to $\sigma = \sqrt{\sigma^2} = \sqrt{1.61} = 1.269$, s is a biased estimator of σ.

4.83 a. $\mu_{\bar{x}} = \mu = 100,\ \sigma_{\bar{x}} = \dfrac{\sigma}{\sqrt{n}} = \dfrac{\sqrt{100}}{\sqrt{4}} = 5$

b. $\mu_{\bar{x}} = \mu = 100,\ \sigma_{\bar{x}} = \dfrac{\sigma}{\sqrt{n}} = \dfrac{\sqrt{100}}{\sqrt{25}} = 2$

c. $\mu_{\bar{x}} = \mu = 100,\ \sigma_{\bar{x}} = \dfrac{\sigma}{\sqrt{n}} = \dfrac{\sqrt{100}}{\sqrt{100}} = 1$

d. $\mu_{\bar{x}} = \mu = 100,\ \sigma_{\bar{x}} = \dfrac{\sigma}{\sqrt{n}} = \dfrac{\sqrt{100}}{\sqrt{50}} = 1.414$

e. $\mu_{\bar{x}} = \mu = 100,\ \sigma_{\bar{x}} = \dfrac{\sigma}{\sqrt{n}} = \dfrac{\sqrt{100}}{\sqrt{500}} = .447$

f. $\mu_{\bar{x}} = \mu = 100,\ \sigma_{\bar{x}} = \dfrac{\sigma}{\sqrt{n}} = \dfrac{\sqrt{100}}{\sqrt{1000}} = .316$

4.85 a. $\mu = \sum xp(x) = 1(.1) + 2(.4) + 3(.4) + 8(.1) = 2.9$

$$\sigma^2 = \sum (x - \mu)^2 p(x) = (1 - 2.9)^2(.1) + (2 - 2.9)^2(.4) + (3 - 2.9)^2(.4) + (8 - 2.9)^2(.1)$$
$$= .361 + .324 + .004 + 2.601 = 3.29$$

$\sigma = \sqrt{3.29} = 1.814$

b. The possible samples, values of $\bar{x}$, and associated probabilities are listed:

Possible Samples	$\bar{x}$	$p(\bar{x})$	Possible Samples	$\bar{x}$	$p(\bar{x})$
1, 1	1	.01	3, 1	2	.04
1, 2	1.5	.04	3, 2	2.5	.16
1, 3	2	.04	3, 3	3	.16
1, 8	4.5	.01	3, 8	5.5	.04
2, 1	1.5	.04	8, 1	4.5	.01
2, 2	2	.16	8, 2	5	.04
2, 3	2.5	.16	8, 3	5.5	.04
2, 8	5	.04	8, 8	8	.01

$P(1, 1) = p(1)p(1) = .1(.1) = .01$
$P(1, 2) = p(1)p(2) = .1(.4) = .04$
$P(1, 3) = p(1)p(3) = .1(.4) = .04$
etc.

The sampling distribution of $\bar{x}$ is:

$\bar{x}$	$p(\bar{x})$
1	.01
1.5	.08
2	.24
2.5	.32
3	.16
4.5	.02
5	.08
5.5	.08
8	.01
	1.00

c. $$\mu_{\bar{x}} = E(\bar{x}) = \sum \bar{x}p(\bar{x}) = 1(.01) + 1.5(.08) + 2(.24) + 2.5(.32) + 3(.16) + 4.5(.02)$$
$$+ 5(.08) + 5.5(.08) + 8(.01)$$
$$= 2.9 = \mu$$

$$\sigma_{\bar{x}}^2 = \sum (\bar{x} - \mu_{\bar{x}})^2 p(\bar{x}) = (1 - 2.9)^2(.01) + (1.5 - 2.9)^2(.08) + (2 - 2.9)^2(.24)$$
$$+ (2.5 - 2.9)^2(.32) + (3 - 2.9)^2(.16) + (4.5 - 2.9)^2(.02)$$
$$+ (5 - 2.9)^2(.08) + (5.5 - 2.9)^2(.08) + (8 - 2.9)^2(.01)$$
$$= .0361 + .1568 + .1944 + .0512 + .0016 + .0512 + .3528$$
$$+ .5408 + .2601$$
$$= 1.645$$

$\sigma_{\bar{x}} = \sqrt{1.645} = 1.283$

$\sigma_{\bar{x}} = \sigma/\sqrt{n} = 1.814/\sqrt{2} = 1.283$

4.87 a. $\mu_{\bar{x}} = \mu = 20,\ \sigma_{\bar{x}} = \sigma/\sqrt{n} = 16/\sqrt{64} = 2$

b. By the Central Limit Theorem, the distribution of $\bar{x}$ is approximately normal. In order for the Central Limit Theorem to apply, n must be sufficiently large. For this problem, $n = 64$ is sufficiently large.

c. $z = \dfrac{\bar{x} - \mu_{\bar{x}}}{\sigma_{\bar{x}}} = \dfrac{15.5 - 20}{2} = -2.25$

d. $z = \dfrac{\bar{x} - \mu_{\bar{x}}}{\sigma_{\bar{x}}} = \dfrac{23 - 20}{2} = 1.50$

4.89 By the Central Limit Theorem, the sampling distribution of $\bar{x}$ is approximately normal with $\mu_{\bar{x}} = \mu = 30$ and $\sigma_{\bar{x}} = \sigma/\sqrt{n} = 16/\sqrt{100} = 1.6$. Using Table III, Appendix A:

a. $P(\bar{x} \geq 28) = P\left(z \geq \dfrac{28 - 30}{1.6}\right) = P(z \geq -1.25) = .5 + .3944 = .8944$

b. $P(22.1 \leq \bar{x} \leq 26.8) = P\left(\dfrac{22.1 - 30}{1.6} \leq z \leq \dfrac{26.8 - 30}{1.6}\right) = P(-4.94 \leq z \leq -2)$
$= .5 - .4772 = .0228$

c. $P(\bar{x} \leq 28.2) = P\left(z \leq \dfrac{28.2 - 30}{1.6}\right) = P(z \leq -1.13) = .5 - .3708 = .1292$

d. $P(\bar{x} \geq 27.0) = P\left(z \geq \dfrac{27.0 - 30}{1.6}\right) = P(z \geq -1.88) = .5 + .4699 = .9699$

4.93 a. $\mu_{\bar{x}} = \mu = 5.1;\ \sigma_{\bar{x}} = \dfrac{\sigma}{\sqrt{n}} = \dfrac{6.1}{\sqrt{150}} = .4981$

b. Because the sample size is large, $n = 150$, the Central Limit Theorem says that the sampling distribution of $\bar{x}$ is approximately normal.

c. $P(\bar{x} > 5.5) = P\left(z > \dfrac{5.5 - 5.1}{.4981}\right) = P(z > .80) = .5 - .2881 = .2119$
(using Table III, Appendix A)

d. $P(4 < \bar{x} < 5) = P\left(\dfrac{4 - 5.1}{.4981} < z < \dfrac{5 - 5.1}{.4981}\right) = P(-2.21 < z < -.20)$
$= .4864 - .0793 = .4071$
(using Table III, Appendix A)

4.95 a. $\mu_{\bar{x}} = \mu = 89.34; \ \sigma_{\bar{x}} = \dfrac{\sigma}{\sqrt{n}} = \dfrac{7.74}{\sqrt{35}} = 1.3083$

b.

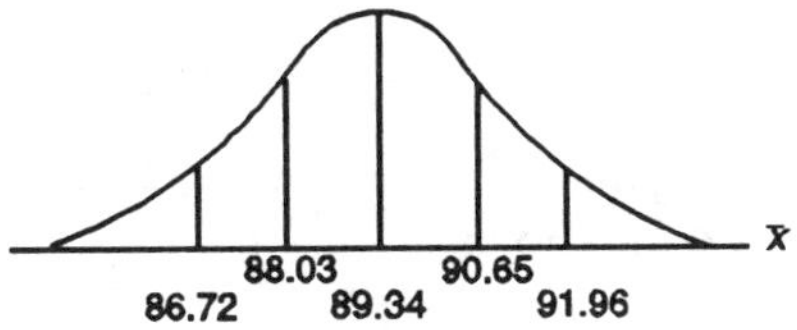

c. $P(\bar{x} > 88) = P\left(z > \dfrac{88 - 89.34}{1.3083}\right) = P(z > -1.02) = .5 + .3461 = .8461$
(using Table III, Appendix A)

d. $P(\bar{x} < 87) = P\left(z < \dfrac{87 - 89.34}{1.3083}\right) = P(z < -1.79) = .5 - .4633 = .0367$
(using Table III, Appendix A)

4.97 a. By the Central Limit Theorem, the distribution of $\bar{x}$ is approximately normal, with $\mu_{\bar{x}} = \mu = 157$ and $\sigma_{\bar{x}} = \sigma/\sqrt{n} = 3/\sqrt{40} = .474$.

The sample mean is 1.3 psi below 157 or $\bar{x} = 157 - 1.3 = 155.7$

$$P(\bar{x} \leq 155.7) = P\left(z \leq \dfrac{155.7 - 157}{.474}\right) = P(z \leq -2.74) = .5 - .4969 = .0031$$
(using Table III, Appendix A)

If the claim is true, it is very unlikely (probability = .0031) to observe a sample mean 1.3 psi below 157 psi. Thus, the actual population mean is probably not 157 but something lower.

b. $P(\bar{x} \leq 155.7) = P\left(z \leq \dfrac{155.7 - 156}{.474}\right) = P(z \leq -.63) = .5 - .2357 = .2643$
(using Table III, Appendix A)

The observed sample is more likely if $\mu = 156$ rather than 157.

$$P(\bar{x} \leq 155.7) = P\left(z \leq \dfrac{155.7 - 158}{.474}\right) = P(z \leq -4.85) = .5 - .5 = 0$$

The observed sample is less likely if $\mu = 158$ rather than 157.

c. $\sigma_{\bar{x}} = \sigma/\sqrt{n} = 2/\sqrt{40} = .316 \qquad \mu_{\bar{x}} = 157$

$$P(\bar{x} \leq 155.7) = P\left(z \leq \dfrac{155.7 - 157}{.316}\right) = P(z \leq -4.11) = .5 - .5 = 0$$
(using Table III, Appendix A)

The observed sample is less likely if $\sigma = 2$ rather than 3.

$\sigma_{\bar{x}} = \sigma/\sqrt{n} = 6/\sqrt{40} = .949 \qquad \mu_{\bar{x}} = 157$

$$P(\bar{x} \leq 155.7) = P\left(z \leq \dfrac{155.7 - 157}{.949}\right) = P(z \leq -1.37) = .5 - .4147 = .0853$$
(using Table III, Appendix A)

The observed sample is more likely if $\sigma = 6$ rather than 3.

4.99 a. The length of time that an exercise physiologist's program takes to elevate her client's heart rate to 140 beats per minute is measured on an interval and thus, is continuous.

b. The number of crimes committed on a college campus per year is a whole number such as 0, 1, 2, etc. This variable is discrete.

c. The number of square feet of vacant office space in a large city is a measurement of area and is measured on an interval. Thus, this variable is continuous.

d. The number of voters who favor a new tax proposal is a whole number such as 0, 1, 2, etc. This variable is discrete.

4.101 a. $P(x = 1) = \begin{pmatrix} 3 \\ 1 \end{pmatrix} (.1)^1(.9)^{3-1} = \frac{3!}{1!2!}(.1)^1(.9)^2 = 3(.1)(.81) = .243$

b. $P(x = 4) = P(x \leq 4) - P(x \leq 3) = .238 - .107 = .131$ from Table II, Appendix A.

c. $P(x = 0) = \begin{pmatrix} 2 \\ 0 \end{pmatrix} (.4)^0(.6)^{2-0} = \frac{2!}{0!2!}(1)(.6)^2 = .36$

d. $P(x = 4) = P(x \leq 4) - P(x \leq 3) = .969 - .812 = .157$ from Table II, Appendix A.

e. $P(x = 12) = P(x \leq 12) - P(x \leq 11) = .184 - .056 = .128$ from Table II, Appendix A.

f. $P(x = 8) = P(x \leq 8) - P(x \leq 7) = .954 - .833 = .121$ from Table II, Appendix A.

4.103 Using Table III, Appendix A:

a. $P(z \leq z_0) = .8708$

$A_1 = .8708 - .5 = .3708$
Looking up area .3708, $z_0 = 1.13$

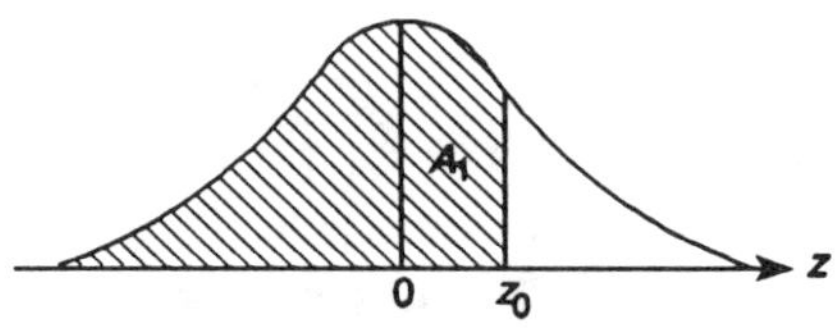

b. $P(z \geq z_o) = .0526$

$A_1 = .5 - .0526 = .4474$
Looking up area .4474, $z_0 = 1.62$

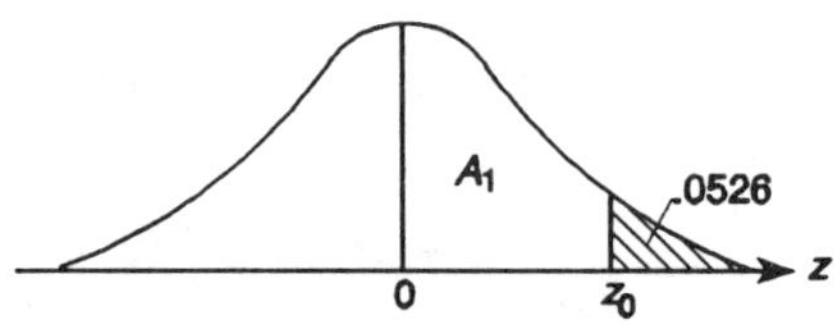

c. $P(z \leq z_0) = .5 \Rightarrow z_0 = 0$

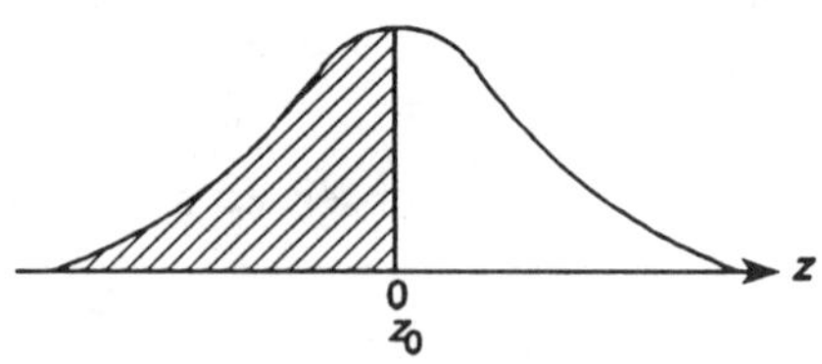

d. $P(-z_0 \leq z \leq z_0) = .8164$

$A_1 = A_2 = .8164/2 = .4082$
Looking up area .4082, $z_0 = 1.33$

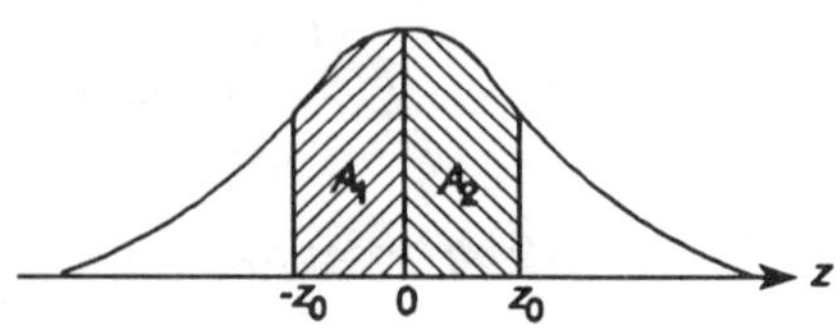

e. $P(z \geq z_0) = .8023$

$A_1 = .8023 - .5 = .3023$
Looking up area .3023, $z = .85$
Since z_0 is to the left of 0, $z_0 = -.85$

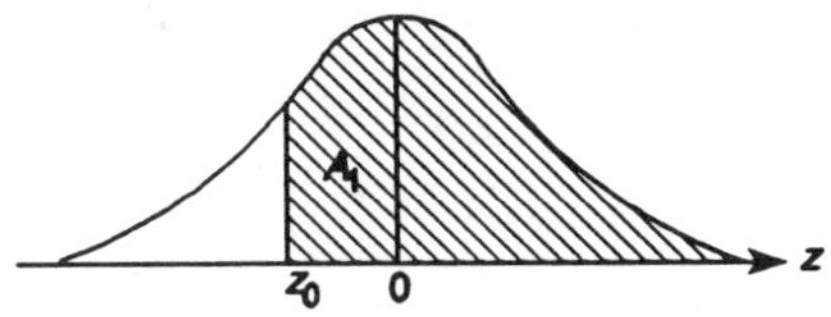

f. $P(z \geq z_0) = .0041$

$A_1 = .5 - .0041 = .4959$
Looking up area .4959, $z_0 = 2.64$

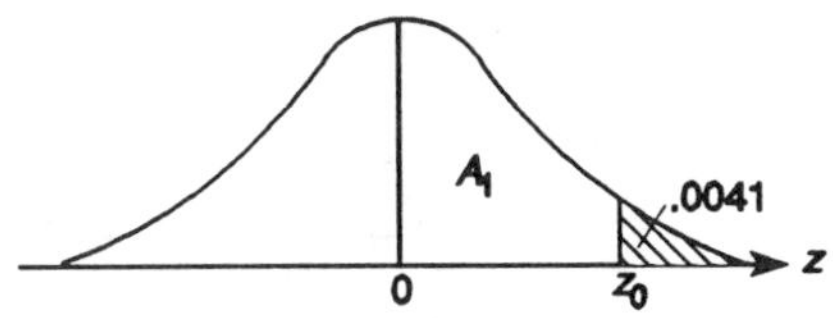

4.105 Using Table III, Appendix A:

a. $P(x \geq x_0) = .5$. Find x_0.

$$P(x \geq x_0) = P\left[z \geq \frac{x_0 - 40}{6}\right] = P(z \geq z_0) = .5 \Rightarrow z_0 = 0$$

$$z_0 = \frac{x_0 - 40}{6} \Rightarrow 0 = \frac{x_0 - 40}{6} \Rightarrow x_0 = 40$$

b. $P(x \leq x_0) = .9911$. Find x_0.

$$P(x \leq x_0) = P\left[z \leq \frac{x_0 - 40}{6}\right] = P(z \leq z_0) = .9911$$

$A_1 = .9911 - .5 = .4911$
Looking up area .4911, $z_0 = 2.37$

$$z_0 = \frac{x_0 - 40}{6} \Rightarrow 2.37 = \frac{x_0 - 40}{6} \Rightarrow x_0 = 54.22$$

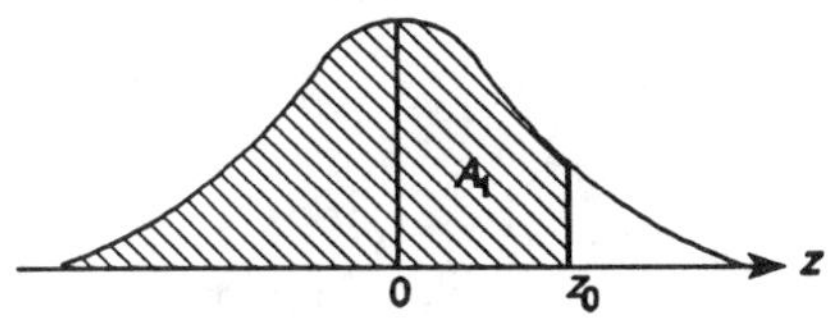

c. $P(x \leq x_0) = .0028$. Find x_0.

$$P(x \leq x_0) = P\left[z \leq \frac{x_0 - 40}{6}\right] = P(z \leq z_0) = .0028$$

$A_1 = .5 - .0028 = .4972$
Looking up area .4972, $z_0 = 2.77$
Since z_0 is to the left of 0, $z_0 = -2.77$

$$z_0 = \frac{x_0 - 40}{6} \Rightarrow -2.77 = \frac{x_0 - 40}{6} \Rightarrow x_0 = 23.38$$

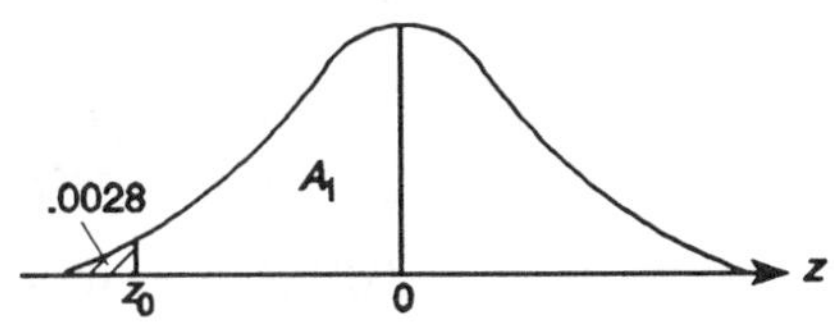

d. $P(x \geq x_0) = .0228$. Find x_0.

$$P(x \geq x_0) = P\left[z \geq \frac{x_0 - 40}{6}\right] = P(z \geq z_0) = .0228$$

$A_1 = .5 - .0228 = .4772$
Looking up area .4772, $z_0 = 2.0$

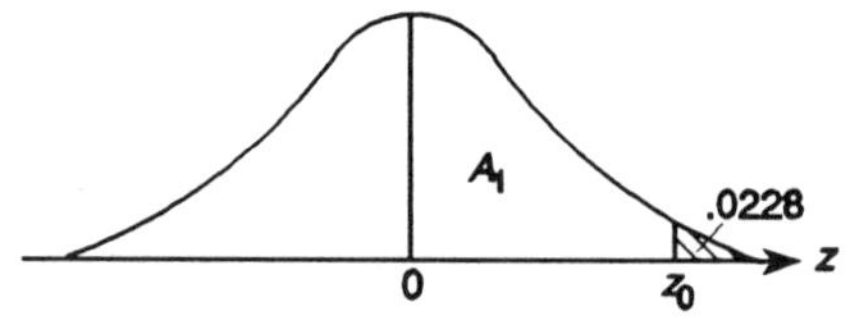

$$z_0 = \frac{x_0 - 40}{6} \Rightarrow 2 = \frac{x_0 - 40}{6} \Rightarrow x_0 = 52$$

e. $P(x \leq x_0) = .1003$. Find x_0.

$$P(x \leq x_0) = P\left[z \leq \frac{x_0 - 60}{8}\right] = P(z \leq z_0) = .1003$$

$A_1 = .5 - .1003 = .3997$
Looking up area .3997, $z = 1.28$.
Since z_0 is to the left of 0, $z_0 = -1.28$

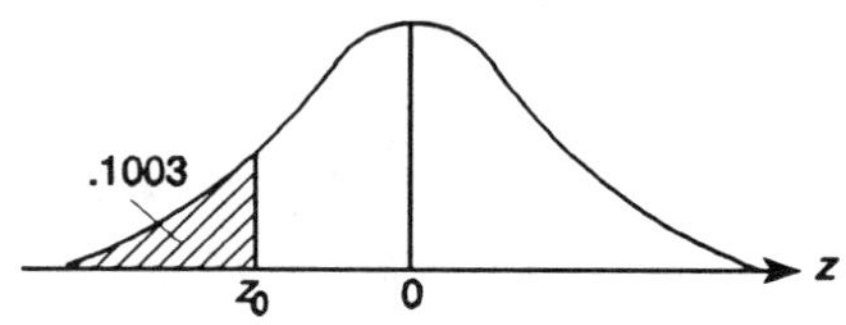

$$z_0 = \frac{x_0 - 40}{6} \Rightarrow -1.28 = \frac{x_0 - 40}{6} \Rightarrow x_0 = 32.32$$

f. $P(x \geq x_0) = .7995$. Find x_0.

$$P(x \geq x_0)\ P\left[z \geq \frac{x_0 - 60}{8}\right] = P(z \geq z_0) = .7995$$

$A_1 = .7995 - .5 = .2995$
Looking up area .2995, $z = .84$.
Since z_0 is to the left of 0, $z_0 = -.84$

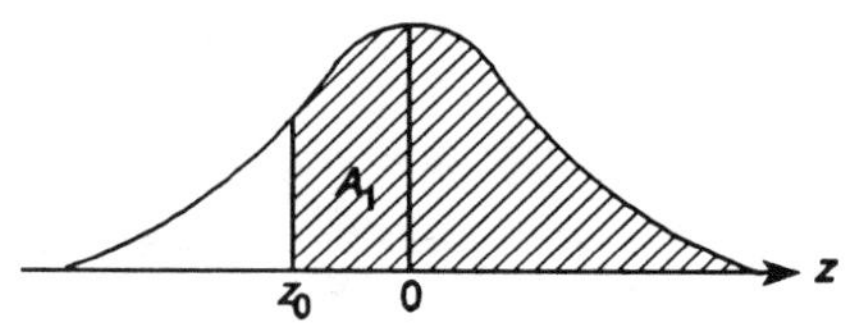

$$z_0 = \frac{x_0 - 40}{6} \Rightarrow -.84 = \frac{x_0 - 40}{6} \Rightarrow x_0 = 34.96$$

4.107 a. $\mu = \sum xp(x) = 10(.2) + 12(.3) + 18(.1) + 20(.4) = 15.4$

$$\sigma^2 = \sum (x - \mu)^2 p(x)$$
$$= (10 - 15.4)^2(.2) + (12 - 15.4)^2(.3) + (18 - 15.4)^2(.1) + (20 - 15.4)^2(.4) = 18.44$$

$$\sigma = \sqrt{18.44} \approx 4.294$$

b. $P(x < 15) = p(10) + p(12) = .2 + .3 = .5$

c. $\mu \pm 2\sigma = 15.4 \pm 2(4.294) \Rightarrow (6.812, 23.988)$

d. $P(6.812 < x < 23.988) = .2 + .3 + .1 + .4 = 1.0$

4.109 By the Central Limit Theorem, the sampling distribution of $\bar{x}$ is approximately normal.

$$\mu_{\bar{x}} = \mu = 19.6,\ \sigma_{\bar{x}} = \frac{3.2}{\sqrt{68}} = .388$$

a. $P(\bar{x} \le 19.6) = P\left[z \le \frac{19.6 - 19.6}{.388}\right] = P(z \le 0) = .5$ (using Table III, Appendix A)

b. $P(\bar{x} \le 19) = P\left[z \le \frac{19 - 19.6}{.388}\right] = P(z \le -1.55) = .5 - .4394 = .0606$
(using Table III, Appendix A)

c. $P(\bar{x} \ge 20.1) = P\left[z \ge \frac{20.1 - 19.6}{.388}\right] = P(z \ge 1.29) = .5 - .4015 = .0985$
(using Table III, Appendix A)

d. $P(19.2 \le \bar{x} \le 20.6) = P\left[\frac{19.2 - 19.6}{.388} \le z \le \frac{20.6 - 19.6}{.388}\right]$
$= P(-1.03 \le z \le 2.58) = .3485 + .4951 = .8436$
(using Table III, Appendix A)

4.111 Given: $\mu = 100$ and $\sigma = 10$

n	1	5	10	20	30	40	50
$\frac{\sigma}{\sqrt{n}}$	10	4.472	3.162	2.236	1.826	1.581	1.414

The graph of $\sigma/\sqrt{n}$ against n is given here:

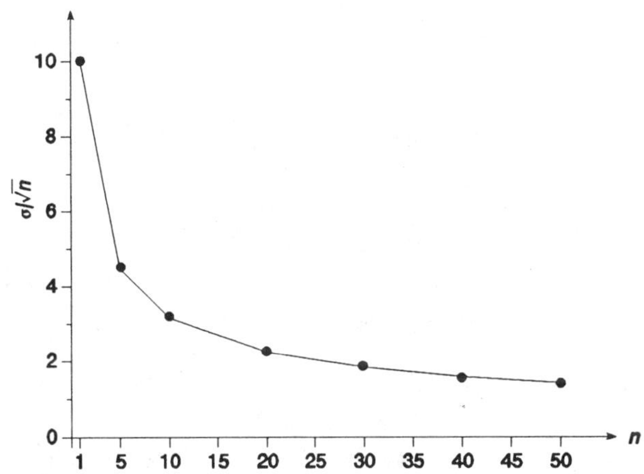

4.113 a. Discrete. There are only 3 possible outcomes for the response: 1, 1.5, or 2.

b. By the Central Limit Theorem, the sampling distribution of $\bar{x}$ is approximately normal ($n = 651$) with $\mu_{\bar{x}} = \mu = 1.50$ and $\sigma_{\bar{x}} = \sigma/\sqrt{n} = .45/\sqrt{651} = .01764$. There are no assumptions necessary.

c. $P(\bar{x} \le 1.45) = P\left[z \le \frac{1.45 - 1.5}{.01764}\right] = P(z \le -2.83) = .5 - .4977 = .0023$
(using Table III, Appendix A)

d. $P(\bar{x} \le 1.36) = P\left[z \le \frac{1.36 - 1.5}{.01764}\right] = P(z \le -7.94) \approx .5 - .5 = 0$
(using Table III, Appendix A)

Since this probability is so small (≈ 0), it is very unlikely that the population mean is 1.50, but probably is less than 1.50.

4.115 Let x = velocity of a galaxy located within the galaxy cluster A2142. Then x is a normal random variable with mean μ = 27,117 km/s and standard deviation σ = 1,280 km/s.

$$P(x \le 24.350) = P\left[z \le \frac{24{,}350 - 27{,}117}{1{,}280}\right] = P(z \le -2.16) = .5 - .4846 = .0154$$

Since the probability of observing a galaxy from the galaxy cluster A2142 with a velocity of 24,350 km/s or slower is so small (p = .0154), it would be very unlikely that the galaxy observed came from the galaxy cluster A2142.

4.117 Let x = score for first time students on mathematics achievement test. Then x is a normal random variable with μ = 77 and σ = 7.3.

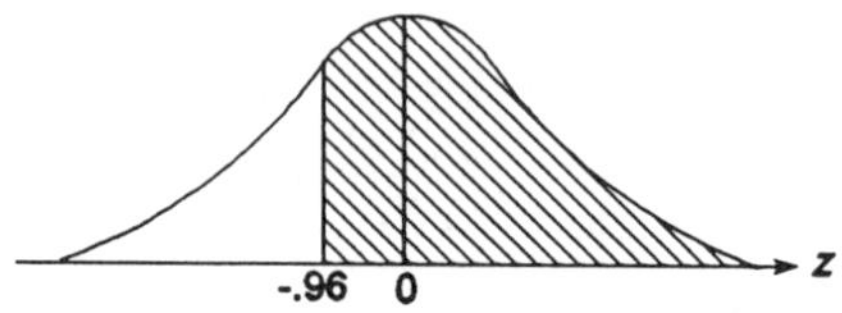

$$P(x \ge 70) = P\left[z \ge \frac{70 - 77}{7.3}\right]$$
$$= P(z \ge -.96)$$
$$= .3315 + .5000 = .8315$$

Thus, 83.15% of students will pass the test the first time.

4.119 a. By the Central Limit Theorem, the sampling distribution of $\bar{x}$ is approximately normal, regardless of the shape of the distribution of the verbal IQ scores. The mean is $\mu_{\bar{x}} = \mu = 107$, and the standard deviation is $\sigma_{\bar{x}} = \sigma/\sqrt{n} = 15/\sqrt{84} = 1.637$.

b. $P(\bar{x} \ge 110) = P\left[z \ge \frac{110 - 107}{1.637}\right] = P(z \ge 1.83) = .5 - .4664 = .0336$
(using Table III, Appendix A)

c. No. If the mean and standard deviation for the nondelinquent juveniles were the same as those for all juveniles, it would be very unlikely (probability = .0336) to observe a sample mean of 110 or higher.

4.121 a. (i) $P(0 < z < 1.2) \approx z(4.4 - z)/10 = 1.2(4.4 - 1.2)/10 = .384$
(ii) $P(0 < z < 2.5) \approx .49$
(iii) $P(z > .8) = .5 - P(0 < z < .8) \approx .5 - .8(4.4 - .8)/10 = .5 - .288 = .212$
(iv) $P(z < 1.0) = .5 + P(0 < z < 1.0) = .5 + 1(4.4 - 1)/10 = .5 + .34 = .84$

b. Using Table III, Appendix A:
(i) $P(0 < z < 1.2) = .3869$
(ii) $P(0 < z < 2.5) = .4938$
(iii) $P(z > .8) = .5 - P(0 < z < .8) = .5 - .2881 = .2119$
(iv) $P(z < 1.0) = .5 + P(0 < z < 1.0) = .5 + .3413 = .8413$

c. For each part, the absolute error is:
(i) Error = $|\,.384 - .3869\,| = .0029$
(ii) Error = $|\,.49 - .4938\,| = .0038$
(iii) Error = $|\,.212 - .2119\,| = .0001$
(iv) Error = $|\,.84 - .8413\,| = .0013$

In all cases,, the absolute error is less than .0052.

4.123 Using Table II with $n = 25$ and $p = .8$:

a. $P(x < 15) = P(x \leq 14) = .006$

b. Since the probability of such an event is so small when $p = .8$, if less than 15 insects die we would conclude that the insecticide is not as effective as claimed.

4.125 By the Central Limit Theorem, the sampling distribution of $\bar{x}$ is approximately normal.

$$\mu_{\bar{x}} = \mu = 75; \ \sigma_{\bar{x}} = \frac{\sigma}{\sqrt{n}} = \frac{10}{\sqrt{36}} = 1.6667$$

a. $$P(\bar{x} > 79) = P\left[z > \frac{79 - 75}{1.6667}\right] = P(z > 2.4) = .5 - .4918 = .0082$$

(using Table III, Appendix A)

b. We must assume the sample size of $n = 36$ is sufficiently large so that the Central Limit Theorem applies.

Inferences Based on a Single Sample: Estimation with Confidence Intervals

Chapter 5

5.1 a. For $\alpha = .10$, $\alpha/2 = .10/2 = .05$. $z_{\alpha/2} = z_{.05}$ is the z-score with .05 of the area to the right of it. The area between 0 and $z_{.05}$ is $.5 - .05 = .4500$. Using Table III, Appendix A, $z_{.05} = 1.645$.

b. For $\alpha = .01$, $\alpha/2 = .01/2 = .005$. $z_{\alpha/2} = z_{.005}$ is the z-score with .005 of the area to the right of it. The area between 0 and $z_{.005}$ is $.5 - .005 = .4950$. Using Table III, Appendix A, $z_{.005} = 2.58$.

c. For $\alpha = .05$, $\alpha/2 = .05/2 = .025$. $z_{\alpha/2} = z_{.025}$ is the z-score with .025 of the area to the right of it. The area between 0 and $z_{.025}$ is $.5 - .025 = .4750$. Using Table III, Appendix A, $z_{.025} = 1.96$.

d. For $\alpha = .20$, $\alpha/2 = .20/2 = .10$. $z_{\alpha/2} = z_{.10}$ is the z-score with .10 of the area to the right of it. The area between 0 and $z_{.10}$ is $.5 - .10 = .4000$. Using Table III, Appendix A, $z_{.10} = 1.28$.

5.3 a. For confidence coefficient .95, $\alpha = .05$ and $\alpha/2 = .05/2 = .025$. From Table III, Appendix A, $z_{.025} = 1.96$. The confidence interval is:

$$\bar{x} \pm z_{.025}\frac{s}{\sqrt{n}} \Rightarrow 25.9 \pm 1.96\frac{2.7}{\sqrt{90}} \Rightarrow 25.9 \pm .56 \Rightarrow (25.34, 26.46)$$

b. For confidence coefficient .90, $\alpha = .10$ and $\alpha/2 = .10/2 = .05$. From Table III, Appendix A, $z_{.05} = 1.645$. The confidence interval is:

$$\bar{x} \pm z_{.05}\frac{s}{\sqrt{n}} \Rightarrow 25.9 \pm 1.645\frac{2.7}{\sqrt{90}} \Rightarrow 25.9 \pm .47 \Rightarrow (25.43, 26.37)$$

c. For confidence coefficient .99, $\alpha = .01$ and $\alpha/2 = .01/2 = .005$. From Table III, Appendix A, $z_{.005} = 2.58$. The confidence interval is:

$$\bar{x} \pm z_{.005}\frac{s}{\sqrt{n}} \Rightarrow 25.9 \pm 2.58\frac{2.7}{\sqrt{90}} \Rightarrow 25.9 \pm .73 \Rightarrow (25.17, 26.63)$$

5.5 a. For confidence coefficient .95, $\alpha = .05$ and $\alpha/2 = .05/2 = .025$. From Table III, Appendix A, $z_{.025} = 1.96$. The confidence interval is:

$$\bar{x} \pm z_{.025}\frac{s}{\sqrt{n}} \Rightarrow 28 \pm 1.96\frac{\sqrt{12}}{\sqrt{75}} \Rightarrow 28 \pm .784 \Rightarrow (27.216, 28.784)$$

b. $$\bar{x} \pm z_{.025}\frac{s}{\sqrt{n}} \Rightarrow 102 \pm 1.96\frac{\sqrt{22}}{\sqrt{200}} \Rightarrow 102 \pm .65 \Rightarrow (101.35, 102.65)$$

c. $\bar{x} \pm z_{.025}\frac{s}{\sqrt{n}} \Rightarrow 15 \pm 1.96\frac{.3}{\sqrt{100}} \Rightarrow 15 \pm .0588 \Rightarrow (14.9412, 15.0588)$

d. $\bar{x} \pm z_{.025}\frac{s}{\sqrt{n}} \Rightarrow 4.05 \pm 1.96\frac{.83}{\sqrt{100}} \Rightarrow 4.05 \pm .163 \Rightarrow (3.887, 4.213)$

e. No. Since the sample size in each part was large (n ranged from 75 to 200), the Central Limit Theorem indicates that the sampling distribution of $\bar{x}$ is approximately normal.

5.7 If we were to repeatedly draw samples of size n from the population and form the interval $\bar{x} \pm 1.96\sigma_{\bar{x}}$ each time, approximately 95% of the intervals would contain μ. We have no way of knowing whether our interval estimate is one of the 95% that contain μ or one of the 5% that do not.

5.9 a. For confidence coefficient .95, $\alpha = .05$ and $\alpha/2 = .05/2 = .025$. From Table III, Appendix A, $z_{.025} = 1.96$. The confidence interval is:

$$\bar{x} \pm z_{.025}\frac{s}{\sqrt{n}} \Rightarrow 33.9 \pm 1.96\frac{3.3}{\sqrt{100}} \Rightarrow 33.9 \pm .647 \Rightarrow (33.253, 34.547)$$

b. $\bar{x} \pm z_{.025}\frac{s}{\sqrt{n}} \Rightarrow 33.9 \pm 1.96\frac{3.3}{\sqrt{400}} \Rightarrow 33.9 \pm .323 \Rightarrow (33.577, 34.223)$

c. For part **a**, the width of the interval is 2(.647) = 1.294. For part **b**, the width of the interval is 2(.323) = .646. When the sample size is quadrupled, the width of the confidence interval is halved.

5.11 a. The interval constructed in Exercise 2.72 was an interval within which 95% of all ***actual observations*** should fall. It is not trying to estimate the mean velocity.

b. For confidence coefficient .95, $\alpha = .05$ and $\alpha/2 = .05/2 = .025$. From Table III, Appendix A, $z_{.025} = 1.96$. The confidence interval is:

$$\bar{x} \pm z_{.025}\frac{s}{\sqrt{n}} \Rightarrow 27{,}117 \pm 1.96\frac{1{,}280}{\sqrt{103}} \Rightarrow 27{,}117 \pm 247.20 \Rightarrow (26{,}869.8, 27{,}364.2)$$

We are 95% confident that the mean light velocity emitted from all galaxies in cluster A2142 is between 26,869.8 and 27,364.2 km/s.

5.13 For confidence coefficient .90, $\alpha = .10$ and $\alpha/2 = .05$. From Table III, Appendix A, $z_{.05} = 1.645$. The confidence interval is:

$$\bar{x} \pm z_{\alpha/2}\frac{s}{\sqrt{n}} \Rightarrow 20.848 \pm 1.645\frac{13.414}{\sqrt{50}} \Rightarrow 20.848 \pm 3.121 \Rightarrow (17.727, 23.969)$$

We are 90% confident the mean attention time given to each pair is between 17.727 and 23.969 hours.

5.15 a. For confidence coefficient .95, $\alpha = .05$ and $\alpha/2 = .025$. From Table III, Appendix A, $z_{.025} = 1.96$. The confidence interval is:

$$\bar{x} \pm z_{\alpha/2}\frac{s}{\sqrt{n}} \Rightarrow 79.73 \pm 1.96\frac{5.96}{\sqrt{30}} \Rightarrow 79.73 \pm 2.133 \Rightarrow (77.597, 81.863)$$

b. We are 95% confident that the mean participation rate for all companies that have 401(k) plans is between 77.597% and 81.863%.

c. We must assume that the sample size ($n = 30$) is sufficiently large so that the Central Limit Theorem applies.

d. Yes. Since 71% is not included in the 95% confidence interval, it can be concluded that this company's participation rate is lower than the population mean.

e. The center of the confidence interval is $\bar{x}$. If 60% is changed to 80%, the value of $\bar{x}$ will increase, thus indicating that the center point will be larger. The value of s^2 will decrease if 60% is replaced by 80%, thus causing the width of the interval to decrease.

5.17 a. For confidence coefficient .95, $\alpha = .05$ and $\alpha/2 = .025$. From Table III, Appendix A, $z_{.025} = 1.96$. The confidence interval is:

$$\bar{x} \pm z_{\alpha/2}\frac{s}{\sqrt{n}}$$

Younger: $4.17 \pm 1.96\frac{.75}{\sqrt{241}} \Rightarrow 4.17 \pm .095 \Rightarrow (4.075, 4.265)$

We are 95% confident that the mean job satisfaction score for all adults in the younger age group is between 4.075 and 4.265.

Middle-Age: $4.04 \pm 1.96\frac{.81}{\sqrt{768}} \Rightarrow 4.04 \pm .057 \Rightarrow (3.983, 4.097)$

We are 95% confident that the mean job satisfaction score for all adults in the middle-age age group is between 3.983 and 4.097.

Older: $4.31 \pm 1.96\frac{.82}{\sqrt{677}} \Rightarrow 4.31 \pm .062 \Rightarrow (4.248, 4.372)$

We are 95% confident that the mean job satisfaction score for all adults in the older age group is between 4.248 and 4.372.

b. Let y = number of 95% confidence intervals that do not contain the population mean in 3 trials. Then y is a binomial random variable with $n = 3$ and $p = .05$.

$$P(y \geq 1) = 1 - P(y = 0) = 1 - \binom{3}{0}.05^0.95^3 = 1 - .857375 = .142625$$

Thus, it is more likely that at least one of three intervals will not contain the population mean than it is for a single confidence interval to miss the population mean.

The probability that a single confidence interval will not contain the population mean is .05.

5.19 a. If x is normally distributed, the sampling distribution of $\bar{x}$ is normal, regardless of the sample size.

b. If nothing is known about the distribution of x, the sampling distribution of $\bar{x}$ is approximately normal if n is sufficiently large. If n is not large, the distribution of $\bar{x}$ is unknown if the distribution of x is not known.

5.21 a. $P(t \geq t_0) = .025$ where df $= 10$

$$t_0 = 2.228$$

b. $P(t \geq t_0) = .01$ where df $= 17$

$$t_0 = 2.567$$

c. $P(t \leq t_0) = .005$ where df $= 6$

Because of symmetry, the statement can be rewritten

$$P(t \geq -t_0) = .005 \text{ where df} = 6$$
$$t_0 = -3.707$$

d. $P(t \leq t_0) = .05$ where df $= 13$

$$t_0 = -1.771$$

5.23 For this sample,

$$\bar{x} = \frac{\sum x}{n} = \frac{1567}{16} = 97.9375$$

$$s^2 = \frac{\sum x^2 - \frac{(\sum x)^2}{n}}{n - 1} = \frac{155{,}867 - \frac{1567^2}{16}}{16 - 1} = 159.9292$$

$$s = \sqrt{s^2} = 12.6463$$

a. For confidence coefficient, .80, $\alpha = 1 - .80 = .20$ and $\alpha/2 = .20/2 = .10$. From Table IV, Appendix A, with df $= n - 1 = 16 - 1 = 15$, $t_{.10} = 1.341$. The 80% confidence interval for μ is:

$$\bar{x} \pm t_{.10}\frac{s}{\sqrt{n}} \Rightarrow 97.94 \pm 1.341\frac{12.6463}{\sqrt{16}} \Rightarrow 97.94 \pm 4.240 \Rightarrow (93.700, 102.180)$$

b. For confidence coefficient, .95, $\alpha = 1 - .95 = .05$ and $\alpha/2 = .05/2 = .025$. From Table IV, Appendix A, with df $= n - 1 = 24 - 1 = 23$, $t_{.025} = 2.131$. The 95% confidence interval for μ is:

$$\bar{x} \pm t_{.025}\frac{s}{\sqrt{n}} \Rightarrow 97.94 \pm 2.131\frac{12.6463}{\sqrt{16}} \Rightarrow 97.94 \pm 6.737 \Rightarrow (91.203, 104.677)$$

The 95% confidence interval for μ is wider than the 80% confidence interval for μ found in part a.

c. For part a:

We are 80% confident that the true population mean lies in the interval 93.700 to 102.180.

For part **b**:

We are 95% confident that the true population mean lies in the interval 91.203 to 104.677.

The 95% confidence interval is wider than the 80% confidence interval because the more confident you want to be that μ lies in an interval, the wider the range of possible values.

5.25 First, we compute the sample mean and standard deviation:

$$\bar{x} = \frac{\sum x}{n} = \frac{604.6}{8} = 75.575$$

$$s^2 = \frac{\sum x^2 - \frac{(\sum x)^2}{n}}{n - 1} = \frac{45{,}835.44 - \frac{604.6^2}{8}}{8 - 1} = 20.3993 \qquad s = \sqrt{20.3993} = 4.5166$$

We must assume that the SPLs are normally distributed.

For confidence coefficient .90, $\alpha = .10$ and $\alpha/2 = .10/2 = .05$. From Table IV, Appendix A, with df $= n - 1 = 8 - 1 = 7$, $t_{.05} = 1.895$. The confidence interval is:

$$\bar{x} \pm t_{.05}\frac{s}{\sqrt{n}} \Rightarrow 75.575 \pm 1.895\frac{4.5166}{\sqrt{8}} \Rightarrow 75.575 \pm 3.026 \Rightarrow (72.549, 78.601)$$

We are 90% confident that the mean SPL of unilateral hearing loss patients when noise is produced at a frequency of 500 hertz is between 72.549 and 78.601 decibels.

5.27 a. For confidence coefficient .99, $\alpha = .01$ and $\alpha/2 = .01/2 = .005$. From Table IV, Appendix A, with df $= n - 1 = 3 - 1 = 2$, $t_{.005} = 9.925$. The confidence interval is:

$$\bar{x} \pm t_{.005}\frac{s}{\sqrt{n}} \Rightarrow 49.3 \pm 9.925\frac{1.5}{\sqrt{3}} \Rightarrow 49.3 \pm 8.60 \Rightarrow (40.70, 57.90)$$

b. We are 99% confident that the mean percentage of B(a)p removed from all soil specimens using the poison is between 40.70% and 57.90%.

c. We must assume that the distribution of the percentages of B(a)p removed from all soil specimens using the poison is normal.

5.29 Some preliminary calculations are:

$$\bar{x} = \frac{\sum x}{n} = \frac{811{,}783}{12} = 67{,}648.583$$

$$s^2 = \frac{\sum x^2 - \frac{(\sum x)^2}{n}}{n - 1} = \frac{58{,}958{,}888{,}430 - \frac{811{,}783^2}{12}}{12 - 1} = 367{,}538{,}046.4$$

$$s = \sqrt{367{,}538{,}046.4} = 19{,}171.2818$$

a. The population is the set of all outstanding principal balances of all home mortgages foreclosed by the bank due to default by the borrower during the last three years. This distribution must be normal.

b. For confidence coefficient .90, $\alpha = .10$ and $\alpha/2 = .05$. From Table IV, Appendix A, with df $= n - 1 = 12 - 1 = 11$, $t_{.05} = 1.796$. The confidence interval is:

$$\bar{x} \pm t_{.05}\frac{s}{\sqrt{n}} \Rightarrow 67{,}648.583 \pm 1.796\left(\frac{19{,}171.2818}{\sqrt{12}}\right)$$
$$\Rightarrow 67{,}648.583 \pm 9939.553 \Rightarrow (\$57{,}709.030, \$77{,}588.136)$$

c. We are 90% confident the mean outstanding principal balance of all home mortgages foreclosed by the bank due to default is between \$57,709.03 and \$77,588.136.

5.31 By the Central Limit Theorem, the sampling distribution of $\hat{p}$ is approximately normal with mean $\mu_{\hat{p}} = p$ and standard deviation $\sigma_{\hat{p}} = \sqrt{\frac{pq}{n}}$.

5.33 a. The sample size is large enough if $\hat{p} \pm 3\sigma_{\hat{p}}$ does not include 0 or 1.

$$\hat{p} \pm 3\sigma_{\hat{p}} \approx .64 \pm 3\sqrt{\frac{.64(.36)}{196}} \Rightarrow .64 \pm .013 \Rightarrow (.537, .743)$$

Since this interval does not contain 0 or 1, the sample size is sufficiently large.

b. For confidence coefficient .95, $\alpha = .05$ and $\alpha/2 = .025$. From Table III, Appendix A, $z_{.025} = 1.96$. The confidence interval is:

$$\hat{p} \pm z_{.025}\sqrt{\frac{\hat{p}\hat{q}}{n}} \Rightarrow .64 \pm 1.96\sqrt{\frac{.64(.36)}{196}} \Rightarrow .64 \pm .067 \Rightarrow (.573, .707)$$

c. We are 95% confident the true value of p is between .372 and .468.

d. "95% confidence" means that if repeated samples of size 196 were selected from the population and 95% confidence intervals formed, 95% of all confidence intervals will contain the true value of p.

5.35 a. Of the 50 observations, 15 like the product $\Rightarrow \hat{p} = \frac{15}{50} = .30$.

To see if the sample size is sufficiently large:

$$\hat{p} \pm 3\sigma_{\hat{p}} \approx \hat{p} \pm 3\sqrt{\frac{\hat{p}\hat{q}}{n}} \Rightarrow .3 \pm 3\sqrt{\frac{.3(.7)}{50}} \Rightarrow .3 \pm .194 \Rightarrow (.106, .494)$$

Since this interval is wholly contained in the interval (0, 1), we may conclude that the normal approximation is reasonable.

For the confidence coefficient .80, $\alpha = .20$ and $\alpha/2 = .10$. From Table III, Appendix A, $z_{.10} = 1.28$. The confidence interval is:

$$\hat{p} \pm z_{.10}\sqrt{\frac{\hat{p}\hat{q}}{n}} \Rightarrow .3 \pm 1.28\sqrt{\frac{.3(.7)}{50}} \Rightarrow .3 \pm .083 \Rightarrow (.217, .383)$$

b. We are 80% confident the proportion of all consumers who like the new snack food is between .217 and .383.

5.37 a. First we must compute $\hat{p}$: $\hat{p} = \frac{x}{n} = \frac{124}{150} = .78$

To see if the sample size is sufficiently large:

$$\hat{p} \pm \sigma_{\hat{p}} \approx \hat{p} \pm 3\sqrt{\frac{\hat{p}\hat{q}}{n}} \Rightarrow .78 \pm 3\sqrt{\frac{.78(.22)}{150}} \Rightarrow .78 \pm .101 \Rightarrow (.679, .881)$$

Since this interval is wholly contained in the interval (0, 1), we may conclude that the normal approximation is reasonable.

For confidence coefficient .90, $\alpha = .10$ and $\alpha/2 = .10/2 = .05$. From Table III, Appendix A, $z_{.05} = 1.645$. The confidence interval is:

$$\hat{p} \pm z_{.05}\sqrt{\frac{pq}{n}} \approx \hat{p} \pm 1.645\sqrt{\frac{\hat{p}\hat{q}}{n}} \Rightarrow .78 \pm 1.645\sqrt{\frac{.78(.22)}{150}} \Rightarrow .78 \pm .056$$
$$\Rightarrow (.724, .836)$$

We are 90% confident that the true proportion of all truck drivers who suffer from sleep apnea is between .724 and .836.

b. Sleep researchers believe that 25% of the population suffer from obstructive sleep apnea. Since the 90% confidence interval for the proportion of truck drivers who suffer from sleep apnea does not contain .25, it appears that the true proportion of truck drivers who suffer from sleep apnea is larger than the proportion of the general population.

5.39 a. First we must compute $\hat{p}$: $\hat{p} = \frac{x}{n} = \frac{58}{90} = .644$

To see if the sample size is sufficiently large:

$$\hat{p} \pm 3\sigma_{\hat{p}} \approx \hat{p} \pm 3\sqrt{\frac{\hat{p}\hat{q}}{n}} \Rightarrow .644 \pm 3\sqrt{\frac{.644(.356)}{90}} \Rightarrow .644 \pm .151 \Rightarrow (.493, .795)$$

Since this interval is wholly contained in the interval (0, 1), we may conclude that the normal approximation is reasonable.

For confidence coefficient .95, $\alpha = .05$ and $\alpha/2 = .05/2 = .025$. From Table III, Appendix A, $z_{.025} = 1.96$. The confidence interval is:

$$\hat{p} \pm z_{.025}\sqrt{\frac{pq}{n}} \approx \hat{p} \pm 1.96\sqrt{\frac{\hat{p}\hat{q}}{n}} \Rightarrow .644 \pm 1.96\sqrt{\frac{.644(.346)}{90}} \Rightarrow .644 \pm .099$$
$$\Rightarrow (.545, .743)$$

We are 90% confident that the true proportion of all subjects who use the bright color level as a cue to being right-side up is between .545 and .743.

b. Since the 95% confidence interval does not contain .50 and both end points are larger than .50, we can conclude that the majority of subjects would select bright color levels over dark color levels as a cue to being right-side up.

5.41 a. First we must compute $\hat{p}$: $\hat{p} = \frac{x}{n} = \frac{97}{185} = .524$

To see if the sample size is sufficiently large:

$$\hat{p} \pm 3\sigma_{\hat{p}} \approx \hat{p} \pm 3\sqrt{\frac{\hat{p}\hat{q}}{n}} \Rightarrow .524 \pm 3\sqrt{\frac{.524(.476)}{185}} \Rightarrow .524 \pm .110 \Rightarrow (.414, .634)$$

Since this interval is wholly contained in the interval (0, 1), we may conclude that the normal approximation is reasonable.

For confidence coefficient .99, $\alpha = .01$ and $\alpha/2 = .01/2 = .005$. From Table III, Appendix A, $z_{.005} = 2.58$. The confidence interval is:

$$\hat{p} \pm z_{.005}\sqrt{\frac{pq}{n}} \approx \hat{p} \pm 2.58\sqrt{\frac{\hat{p}\hat{q}}{n}} \Rightarrow .524 \pm 2.58\sqrt{\frac{.524(.476)}{185}} \Rightarrow .524 \pm .095$$
$$\Rightarrow (.429, .619)$$

b. We are 99% confident that the true proportion of bottlenose dolphin signature whistles that are Type "a" whistles is between .429 and .619.

5.43 The point estimate for the proportion of all youths who think drugs is their greatest problem is:

$$\hat{p} = \frac{x}{n} = \frac{128}{400} = .32$$

The "margin of error" used by the media is 2 standard deviations above and below the point estimate. For this estimate, the "margin of error" is:

$$2\sigma_{\hat{p}} \approx 2\sqrt{\frac{\hat{p}\hat{q}}{n}} \Rightarrow 2\sqrt{\frac{.32(.68)}{400}} \Rightarrow .047$$

Thus, the "margin of error" is $\pm .047$.

The point estimate for the proportion of all youths who think crime in school is their greatest problem is:

$$\hat{p} = \frac{x}{n} = \frac{62}{400} = .155$$

The "margin of error" used by the media is 2 standard deviations above and below the point estimate. For this estimate, the "margin of error" is:

$$2\sigma_{\hat{p}} \approx 2\sqrt{\frac{\hat{p}\hat{q}}{n}} \Rightarrow 2\sqrt{\frac{.155(.845)}{400}} \Rightarrow .036$$

Thus, the "margin of error" is $\pm .036$.

5.45 To compute the necessary sample size, use

$$n = \frac{(z_{\alpha/2})^2\sigma^2}{B^2} \text{ where } \alpha = 1 - .95 = .05 \text{ and } \alpha/2 = .05/2 = .025$$

From Table III, Appendix A, $z_{.025} = 1.96$. Thus,

$$n = \frac{(1.96)^2(5.4)}{.2^2} = 518.616 \approx 519$$

You would need to take 519 samples.

5.47 a. Range = 39 − 31 = 8. $\sigma \approx \frac{\text{Range}}{4} = \frac{8}{4} = 2$

For confidence coefficient .90, α = .10 and $\alpha/2$ = .05. From Table III, Appendix A, $z_{.05}$ = 1.645.

The sample size is $n = \frac{z_{\alpha/2}^2 \sigma^2}{B^2} = \frac{1.645^2(2^2)}{.15^2} = 481.07 \approx 482$

b. $\sigma \approx \frac{\text{Range}}{6} = \frac{8}{6} = 1.333$

The sample size is $n = \frac{z_{\alpha/2}^2 \sigma^2}{B^2} = \frac{1.645^2(1.333^2)}{.15^2} = 213.7 \approx 214$

5.49 For confidence coefficient .90, α = .10 and $\alpha/2$ = .05. From Table III, Appendix A, $z_{.05}$ = 1.645.

We know $\hat{p}$ is in the middle of the interval, so $\hat{p} = \frac{.54 + .26}{2} = .4$

The confidence interval is $\hat{p} \pm z_{.05}\sqrt{\frac{\hat{p}\hat{q}}{n}} \Rightarrow .4 \pm 1.645\sqrt{\frac{.4(.6)}{n}}$

We know $.4 - 1.645\sqrt{\frac{.4(.6)}{n}} = .26$

$\Rightarrow .4 - \frac{.8059}{\sqrt{n}} = .26$

$\Rightarrow .4 - .26 = \frac{.8059}{\sqrt{n}} \Rightarrow \sqrt{n} = \frac{.8059}{.14} = 5.756$

$\Rightarrow n = 5.756^2 = 33.1 \approx 34$

5.51 a. The width of a confidence interval is $2B = 2z_{\alpha/2}\frac{\sigma}{\sqrt{n}}$

For confidence coefficient .95, α = 1 − .95 = .05 and $\alpha/2$ = .05/2 = .025. From Table III, Appendix A, $z_{.025}$ = 1.96.

For n = 16,

$$W = 2z_{\alpha/2}\frac{\sigma}{\sqrt{n}} = 2(1.96)\frac{1}{\sqrt{16}} = 0.98$$

For n = 25,

$$W = 2z_{\alpha/2}\frac{\sigma}{\sqrt{n}} = 2(1.96)\frac{1}{\sqrt{25}} = 0.784$$

For n = 49,

$$W = 2z_{\alpha/2}\frac{\sigma}{\sqrt{n}} = 2(1.96)\frac{1}{\sqrt{49}} = 0.56$$

For $n = 100$,

$$W = 2z_{\alpha/2}\frac{\sigma}{\sqrt{n}} = 2(1.96)\frac{1}{\sqrt{100}} = 0.392$$

For $n = 400$,

$$W = 2z_{\alpha/2}\frac{\sigma}{\sqrt{n}} = 2(1.96)\frac{1}{\sqrt{400}} = 0.196$$

b.

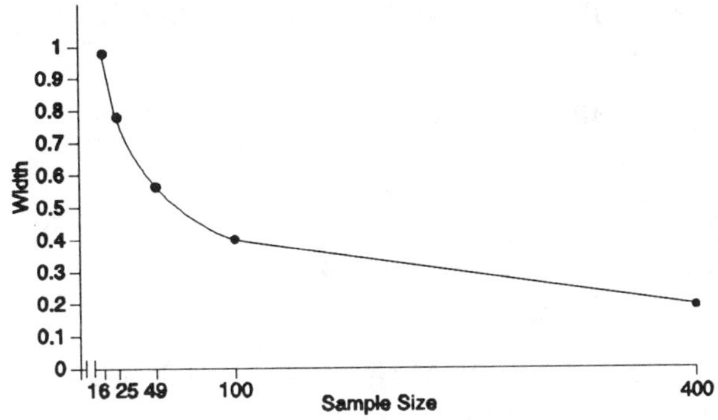

5.53 For confidence coefficient .95, $\alpha = .05$ and $\alpha/2 = .05/2 = .025$. From Table III, Appendix A, $z_{.025} = 1.96$. For this study,

$\hat{p} = \frac{190}{532} = .36$. Using $\hat{p}$ to estimate p, we get

$$n = \frac{(z_{\alpha/2})^2 pq}{B^2} \approx \frac{1.96^2(.36)(.64)}{.03^2} = 983.4 \approx 984$$

No. The sample size needed to be 984 instead of 532.

5.55 For confidence coefficient .95, $\alpha = .05$ and $\alpha/2 = .025$. From Table III, Appendix A, $z_{.025} = 1.96$.

The sample size is $n = \frac{z_{\alpha/2}^2\sigma^2}{B^2} = \frac{1.96^2(.75^2)}{.04^2} = 1350.56 \approx 1351$

5.57 For confidence coefficient .90, $\alpha = .10$ and $\alpha/2 = .05$. From Table III, Appendix A, $z_{.05} = 1.645$.

The sample size is $n = \frac{(z_{\alpha/2})^2\sigma^2}{B^2} = \frac{(1.645)^2(10)^2}{1^2} = 270.6 \approx 271$

5.59 For confidence coefficient .95, $\alpha = .05$ and $\alpha/2 = .025$. From Table III, Appendix A, $z_{.025} = 1.96$.

The sample size is $n = \frac{(z_{\alpha/2})^2 pq}{B^2} = \frac{(1.96)^2(.05)(.95)}{.02^2} = 456.19 \approx 457$

NHSA needs to sample $n = 457$ tires to meet the desired specifications.

At \$25 a tire, 25(457) = \$11,425 is needed. Since NHSA only has \$10,000, it cannot attain its goal while staying within the budget.

5.61 For confidence coefficient .95, α = .05 and $\alpha/2$ = .025. From Table III, Appendix A, $z_{.025}$ = 1.96. We will assume that the true proportion is approximately .108 from Exercise 5.38.

The sample size is $n = \frac{(z_{\alpha/2})^2 pq}{B^2} = \frac{(1.96)^2(.108)(.892)}{.01^2} = 3700.8 \approx 3701$

You would need to take n = 3,701 samples.

5.63 a. For confidence coefficient .99, α = .01 and $\alpha/2$ = .005. From Table III, Appendix A, $z_{.005}$ = 2.58. The confidence interval is:

$$\bar{x} \pm z_{\alpha/2}\frac{s}{\sqrt{n}} \Rightarrow 32.5 \pm 2.58\frac{30}{\sqrt{225}} \Rightarrow 32.5 \pm 5.16 \Rightarrow (27.34,\ 37.66)$$

b. The sample size is $n = \frac{(z_{\alpha/2})^2\sigma^2}{B^2} = \frac{2.58^2(30)^2}{.5^2} = 23{,}963.04 \approx 23{,}964$

c. "99% confidence" means that if repeated samples of size 225 were selected from the population and 99% confidence intervals constructed for the population mean, then 99% of all the intervals constructed will contain the population mean.

5.65 The parameters of interest for the problems are:

(1) The question requires a categorical response. One parameter of interest might be the proportion, p, of all Americans over 18 years of age who think their health is generally very good or excellent.
(2) A parameter of interest might be the mean number of days, μ, in the previous 30 days that all Americans over 18 years of age felt that their physical health was not good because of injury or illness.
(3) A parameter of interest might be the mean number of days, μ, in the previous 30 days that all Americans over 18 years of age felt that their mental health was not good because of stress, depression, or problems with emotions.
(4) A parameter of interest might be the mean number of days, μ, in the previous 30 days that all Americans over 18 years of age felt that their physical or mental health prevented them from performing their usual activities.

5.67 For confidence coefficient .95, α = .05 and $\alpha/2$ = .025. From Table III, Appendix A, $z_{.025}$ = 1.96. The confidence interval is:

$$\bar{x} \pm z_{.025}\frac{s}{\sqrt{n}} \Rightarrow .812 \pm 1.96\frac{1.50476}{\sqrt{52}} \Rightarrow .812 \pm .409 \Rightarrow (.403,\ 1.221)$$

We are 95% confident that the mean solution time for the hybrid algorithm is between .403 and 1.221 seconds.

5.69 a. First we must compute $\hat{p}$: $\hat{p} = \frac{x}{n} = \frac{11{,}239}{43{,}732} = .257$

To see if the sample size is sufficiently large:

$$\hat{p} \pm 3\sigma_{\hat{p}} \approx \hat{p} \pm 3\sqrt{\frac{\hat{p}\hat{q}}{n}} \Rightarrow .257 \pm 3\sqrt{\frac{.257(.743)}{43{,}732}} \Rightarrow .257 \pm .006 \Rightarrow (.251,\ .263)$$

Since this interval is wholly contained in the interval (0, 1), we may conclude that the normal approximation is reasonable.

For confidence coefficient .90, $\alpha = .10$ and $\alpha/2 = .10/2 = .05$. From Table III, Appendix A, $z_{.05} = 1.645$. The confidence interval is:

$$\hat{p} \pm z_{.05}\sqrt{\frac{pq}{n}} \approx \hat{p} \pm 1.645\sqrt{\frac{\hat{p}\hat{q}}{n}} \Rightarrow .257 \pm 1.645\sqrt{\frac{.257(.743)}{43{,}732}} \Rightarrow .257 \pm .003$$
$$\Rightarrow (.254, .260)$$

We are 99% confident that the true percentage of U.S. adults who currently smoke is between 25.4% and 26.0%.

b. First we must compute $\hat{p}$: $\hat{p} = \frac{x}{n} = \frac{10{,}539}{43{,}732} = .241$

To see if the sample size is sufficiently large:

$$\hat{p} \pm 3\sigma_{\hat{p}} \approx \hat{p} \pm 3\sqrt{\frac{\hat{p}\hat{q}}{n}} \Rightarrow .241 \pm 3\sqrt{\frac{.241(.759)}{43{,}732}} \Rightarrow .241 \pm .006 \Rightarrow (.235, .247)$$

Since this interval is wholly contained in the interval (0, 1), we may conclude that the normal approximation is reasonable.

For confidence coefficient .90, $\alpha = .10$ and $\alpha/2 = .10/2 = .05$. From Table III, Appendix A, $z_{.05} = 1.645$. The confidence interval is:

$$\hat{p} \pm z_{.05}\sqrt{\frac{pq}{n}} \approx \hat{p} \pm 1.645\sqrt{\frac{\hat{p}\hat{q}}{n}} \Rightarrow .241 \pm 1.645\sqrt{\frac{.241(.759)}{43{,}732}} \Rightarrow .241 \pm .003$$
$$\Rightarrow (.238, .244)$$

We are 99% confident that the true percentage of U.S. adults who are former smokers is between 23.8% and 24.4%.

5.71 First we compute the sample proportion:

$$\hat{p} = \frac{x}{n} = \frac{172}{323} = .5325$$

We must check to see if the sample size is sufficiently large:

$$\hat{p} \pm 3\sigma_{\hat{p}} \approx \hat{p} \pm 3\sqrt{\frac{\hat{p}\hat{q}}{n}} \Rightarrow .5325 \pm 3\sqrt{\frac{.5325(.4675)}{323}}$$
$$\Rightarrow .5325 \pm .0833 \Rightarrow (.4492, .6158)$$

Since the interval is wholly contained in the interval (0, 1), we may conclude that the normal approximation is reasonable.

For confidence coefficient .90, $\alpha = .10$ and $\alpha/2 = .05$. From Table III, Appendix A, $z_{.05} = 1.645$. The confidence interval is:

$$\hat{p} \pm z_{.05}\sqrt{\frac{\hat{p}\hat{q}}{n}} \Rightarrow .5325 \pm 1.645\sqrt{\frac{.5325(.4675)}{323}} \Rightarrow .5325 \pm .0457 \Rightarrow (.4868, .5782)$$

We are 90% confident that the true proportion of post-retirement travelers who stay between 4 and 7 nights on a typical trip is between .4868 and .5782.

5.73 First we must compute $\hat{p}$: $\hat{p} = \frac{x}{n} = \frac{10}{80} = .125$

For confidence coefficient .99, $\alpha = .01$ and $\alpha/2 = .005$. From Table III, Appendix A, $z_{.005} = 2.58$. The confidence interval is:

$$\hat{p} \pm z_{\alpha/2}\sqrt{\frac{\hat{p}\hat{q}}{n}} \Rightarrow .125 \pm 2.58\sqrt{\frac{(.125)(.875)}{80}} \Rightarrow .125 \pm .095 \Rightarrow (.030,\ .220)$$

We are 99% confident that the true proportion of starlings that will be killed by the substance is between .030 and .220.

5.75 For confidence coefficient .90, $\alpha = .10$ and $\alpha/2 = .05$. From Table III, Appendix A, $z_{.05} = 1.645$. The confidence interval is:

$$\bar{x} \pm z_{\alpha/2}\frac{s}{\sqrt{n}} \Rightarrow 2 \pm 1.645\frac{1.7}{\sqrt{35}} \Rightarrow 2 \pm .47 \Rightarrow (1.53,\ 2.47)$$

We are 90% confident the mean number of cavities per child under age of 12 who lives in the sampled environment is between 1.53 and 2.47.

5.77 a. The point estimate for the mean wear is $\bar{x} = 42{,}250$ miles.

b. For confidence coefficient .90, $\alpha = 1 - .90 = .10$ and $\alpha/2 = .10/2 = .05$. From Table IV, Appendix A, with df $= n - 1 = 20 - 1 = 19$, $t_{.05} = 1.729$. The 90% confidence interval is:

$$\bar{x} \pm t_{.05}\frac{s}{\sqrt{n}} \Rightarrow 42{,}250 \pm 1.729\frac{4355}{\sqrt{20}} \Rightarrow 42{,}250 \pm 1{,}683.713$$
$$\Rightarrow (40{,}566.287,\ 43{,}933.713)$$

c. Interval estimation is better. The chance of estimating the true mean exactly with a single number is 0. With a confidence interval, we can estimate the true mean using a range of values. We are then fairly confident the true mean will fall in this range.

5.79 a. The point estimate for the fraction of the entire market who refuse to purchase bars is:

$$\hat{p} = \frac{x}{n} = \frac{23}{244} = .094$$

b. To see if the sample size is sufficient:

$$\hat{p} \pm 3\sqrt{\frac{\hat{p}\hat{q}}{n}} \Rightarrow .094 \pm 3\sqrt{\frac{(.094)(.906)}{244}} \Rightarrow .094 \pm .056 \Rightarrow (.038,\ .150)$$

Since the interval above is contained in the interval (0, 1), the sample size is sufficiently large.

c. For confidence coefficient .95, $\alpha = 1 - .95 = .05$ and $\alpha/2 = .05/2 = .025$. From Table III, Appendix A, $z_{.025} = 1.96$. The confidence interval is:

$$\hat{p} \pm z_{.025}\sqrt{\frac{\hat{p}\hat{q}}{n}} \Rightarrow .094 \pm 1.96\sqrt{\frac{.094(.906)}{244}} \Rightarrow .094 \pm .037 \Rightarrow (.057,\ .131)$$

d. The best estimate of the true fraction of the entire market who refuse to purchase bars six months after the poisoning is .094. We are 95% confident the true fraction of the entire market who refuse to purchase bars six months after the poisoning is between .057 and .131.

Inferences Based on a Single Sample: Tests of Hypothesis

Chapter 6

6.1 The null hypothesis is the "status quo" hypothesis, while the alternative hypothesis is the research hypothesis.

6.3 The "level of significance" of a test is α. This is the probability that the test statistic will fall in the rejection region when the null hypothesis is true.

6.5 The four possible results are:

1. Rejecting the null hypothesis when it is true. This would be a Type I error.
2. Accepting the null hypothesis when it is true. This would be a correct decision.
3. Rejecting the null hypothesis when it is false. This would be a correct decision.
4. Accepting the null hypothesis when it is false. This would be a Type II error.

6.7 When you reject the null hypothesis in favor of the alternative hypothesis, this does not prove the alternative hypothesis is correct. We are $100(1 - \alpha)\%$ confident that there is sufficient evidence to conclude that the alternative hypothesis is correct.

If we were to repeatedly draw samples from the population and perform the test each time, approximately $100(1 - \alpha)\%$ of the tests performed would yield the correct decision.

6.9 a. Since the company must give proof the drug is safe, the null hypothesis would be the drug is unsafe. The alternative hypothesis would be the drug is safe.

b. A Type I error would be concluding the drug is safe when it is not safe. A Type II error would be concluding the drug is not safe when it is. α is the probability of concluding the drug is safe when it is not. β is the probability of concluding the drug is not safe when it is.

c. In this problem, it would be more important for α to be small. We would want the probability of concluding the drug is safe when it is not to be as small as possible.

6.11 a. Probability of Type I error
$= P(z > 1.96) = .5 - .4750 = .0250$
(from Table III, Appendix A)

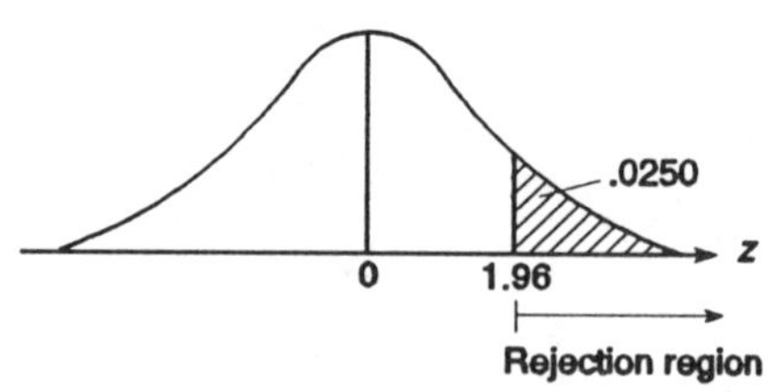

b. Probability of Type I error
$= P(z > 1.645) = .5 - .4500 = .05$
(from Table III, Appendix A)

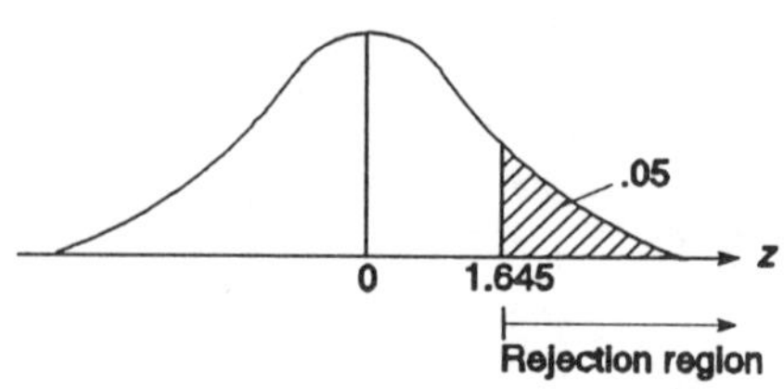

c. Probability of Type I error
$= P(z > 2.58) = .5 - .4951 = .0049$
(from Table III, Appendix A)

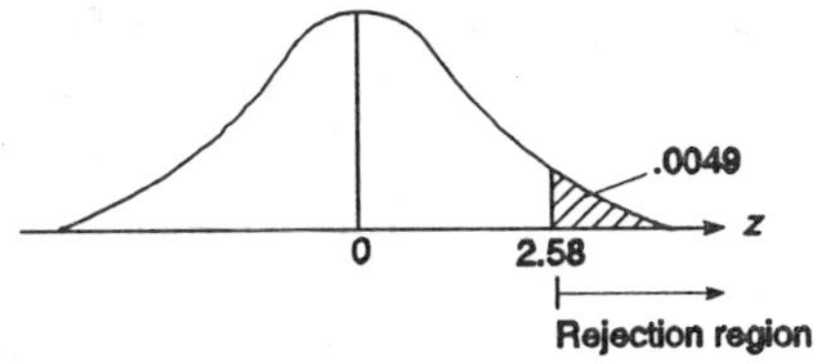

d. Probability of Type I error
$= P(z < -1.28) = .5 - .3997 = .1003$
(from Table III, Appendix A)

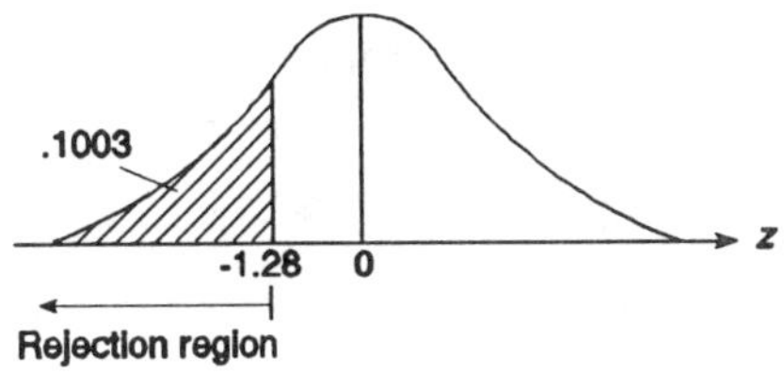

e. Probability of Type I error
$= P(z < -1.645) + P(z > 1.645)$
$= .5 - .4500 + .5 - .4500 = .05 + .05 = .10$
(from Table III, Appendix A)

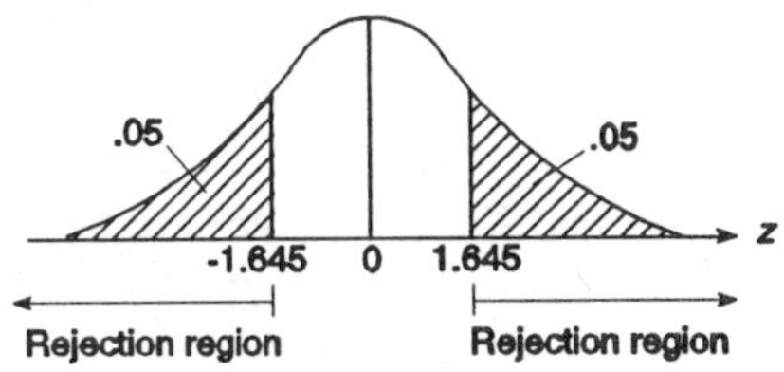

f. Probability of Type I error
$= P(z < -2.58) + P(z > 2.58)$
$= .5 - .4951 + .5 - .4951 = .0049 + .0049$
$= .0098$ (from Table III, Appendix A)

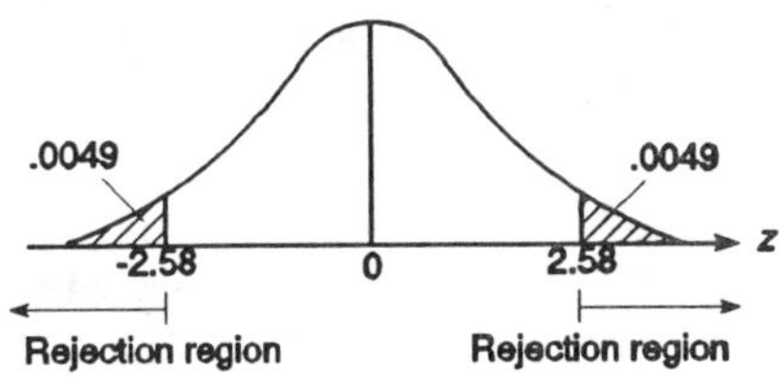

6.13 a. H_0: $\mu = 100$
H_a: $\mu > 100$

The test statistic is $z = \dfrac{\bar{x} - \mu_0}{\sigma_{\bar{x}}} = \dfrac{\bar{x} - \mu_0}{\sigma/\sqrt{n}} = \dfrac{110 - 100}{60/\sqrt{100}} = 1.67$

The rejection region requires $\alpha = .05$ in the upper tail of the z-distribution. From Table III, Appendix A, $z_{.05} = 1.645$. The rejection region is $z > 1.645$.

Since the observed value of the test statistic falls in the rejection region, ($z = 1.67 > 1.645$), H_0 is rejected. There is sufficient evidence to indicate the true population mean is greater than 100 at $\alpha = .05$.

b. H_0: $\mu = 100$
H_a: $\mu \neq 100$

The test statistic is $z = \dfrac{\bar{x} - \mu_0}{\sigma_{\bar{x}}} = \dfrac{110 - 100}{60/\sqrt{100}} = 1.67$

The rejection region requires $\alpha/2 = .05/2 = .025$ in each tail of the z-distribution. From Table III, Appendix A, $z_{.025} = 1.96$. The rejection region is $z < -1.96$ or $z > 1.96$.

Since the observed value of the test statistic does not fall in the rejection region, ($z = 1.67 \ngtr 1.96$), H_0 is not rejected. There is insufficient evidence to indicate μ does not equal 100 at $\alpha = .05$.

c. In part **a**, we rejected H_0 and concluded the mean was greater than 100. In part **b**, we did not reject H_0. There was insufficient evidence to conclude the mean was different from 100. Because the alternative hypothesis in part **a** is more specific than the one in **b**, it is easier to reject H_0.

6.15 a. To determine whether the true mean PTSD score of all World War II aviator POWs is less than 16, we test:

$$H_0: \mu = 16$$
$$H_a: \mu < 16$$

b. The test statistic is $z = \dfrac{\bar{x} - \mu_0}{\sigma_{\bar{x}}} = \dfrac{9 - 16}{9.32/\sqrt{33}} = -4.31$

The rejection region requires $\alpha = .10$ in the lower tail of the z-distribution. From Table III, Appendix A, $z_{.10} = 1.28$. The rejection region is $z < -1.28$.

Since the observed value of the test statistic falls in the rejection region ($z = -4.31 < -1.28$), H_0 is rejected. There is sufficient evidence to indicate that the true mean PTSD score of all World War II aviator POWs is less than 16 at $\alpha = .10$.

The practical implications of the test are that the World War II aviator POWs have a lower level PTSD level on the average than the POWs from Vietnam.

c. The sample used in this study was a self-selected sample—only 33 of the 239 located survivors responded. Very often, self-selected respondents are not representative of the population. Here, those former POWs who are more comfortable with their lives may be more willing to respond than those who are less comfortable. Those who are less comfortable may be suffering more from PTSD than those who are more comfortable. Also, it may not be fair to compare the survivors from World War II to the survivors of Vietnam. The World War II survivors are more removed from their imprisonment than those from the Vietnam war. Also, many of the World War II POWs probably are no longer living. Again, those still alive may be the ones who are more comfortable with their lives.

6.17 a. To determine if the population mean ratio of all bones of this particular species differs from 8.5, we test:

$$H_0: \mu = 8.5$$
$$H_a: \mu \neq 8.5$$

The test statistic is $z = \dfrac{\bar{x} - \mu_0}{\sigma_{\bar{x}}} = \dfrac{9.258 - 8.5}{1.204/\sqrt{41}} = 4.03$

The rejection region requires $\alpha/2 = .01/2 = .005$ in each tail of the z-distribution. From Table III, Appendix A, $z_{.005} = 2.58$. The rejection region is $z < -2.58$ or $z > 2.58$.

Since the observed value of the test statistic falls in the rejection region ($z = 4.03 > 2.58$), H_0 is rejected. There is sufficient evidence to indicate that the true population mean ratio of all bones of this particular species differs from 8.5 at $\alpha = .01$.

b. The practical implications of the test in part **a** is that the species from which these particular bones came from is probably not species A.

6.19 a. To determine if the sample data refute the manufacturer's claim, we test:

H_0: $\mu = 10$
H_a: $\mu < 10$

b. A Type I error is concluding the mean number of solder joints inspected per second is less than 10 when, in fact, it is 10 or more.

A Type II error is concluding the mean number of solder joints inspected per second is at least 10 when, in fact, it is less than 10.

c. H_0: $\mu = 10$
H_a: $\mu < 10$

The test statistic is $z = \dfrac{\bar{x} - \mu_0}{\sigma_{\bar{x}}} = \dfrac{9.29 - 10}{2.10/\sqrt{48}} = -2.34$

The rejection region requires $\alpha = .05$ in the lower tail of the z-distribution. From Table III, Appendix A, $z_{.05} = 1.645$. The rejection region is $z < -1.645$.

Since the observed value of the test statistic falls in the rejection region ($z = -2.34 < -1.645$), H_0 is rejected. There is sufficient evidence to indicate the mean number of inspections per second is less than 10 at $\alpha = .05$.

6.21 a. To determine if the true mean inbreeding coefficient μ for this species of wasp exceeds 0, we test:

H_0: $\mu = 0$
H_a: $\mu > 0$

The test statistic is $z = \dfrac{\bar{x} - \mu_0}{\sigma_{\bar{x}}} = \dfrac{.044 - 0}{.884/\sqrt{197}} = 0.70$

The rejection region requires $\alpha = .05$ in the upper tail of the z-distribution. From Table III, Appendix A, $z_{.05} = 1.645$. The rejection region is $z > 1.645$.

Since the observed value of the test statistic does not fall in the rejection region ($z = 0.70 \ngtr 1.645$), H_0 is not rejected. There is insufficient evidence to indicate that the true mean inbreeding coefficient μ for this species of wasp exceeds 0 at $\alpha = .05$.

b. This result agrees with that of Exercise 5.16. The confidence interval in Exercise 5.16 was $(-.118, .206)$. Since this interval contains 0, there is no evidence to indicate that the mean is different from 0. This is the same conclusion that was reached in part a.

6.23 a. Since the p-value $= .10$ is greater than $\alpha = .05$, H_0 is not rejected.

b. Since the p-value $= .05$ is less than $\alpha = .10$, H_0 is rejected.

c. Since the p-value $= .001$ is less than $\alpha = .01$, H_0 is rejected.

d. Since the p-value $= .05$ is greater than $\alpha = .025$, H_0 is not rejected.

e. Since the p-value $= .45$ is greater than $\alpha = .10$, H_0 is not rejected.

6.25 p-value $= P(z \geq 2.17) = .5 - P(0 < z < 2.17) = .5 - .4850 = .0150$

(using Table III, Appendix A)

6.27 $z = \dfrac{\bar{x} - \mu_0}{\sigma_{\bar{x}}} = \dfrac{49.4 - 50}{4.1/\sqrt{100}} = -1.46$

p-value $= P(z \geq -1.46) = .5 + .4279 = .9279$

There is no evidence to reject H_0 for $\alpha \leq .10$.

6.29 a. The p-value reported by SAS is for a two-tailed test. Thus, $P(z \leq -1.63) + P(z \geq 1.63) = .1032$. For this one-tailed test, the p-value $= P(z \leq -1.63) = .1032/2 = .0516$.

Since the p-value $= .0516 > \alpha = .05$, H_0 is not rejected. There is insufficient evidence to indicate $\mu < 75$ at $\alpha = .05$.

b. For this one-tailed test, the p-value $= P(z \leq 1.63)$. Since $P(z \leq -1.63) = .1032/2 = .0516$, $P(z \leq 1.63) = 1 - .0516 = .9484$.

Since the p-value $= .9484 > \alpha = .10$, H_0 is not rejected. There is insufficient evidence to indicate $\mu < 75$ at $\alpha = .10$.

c. For this one-tailed test, the p-value $= P(z \geq 1.63) = .1032/2 = .0516$.

Since the p-value $= .0516 < \alpha = .10$, H_0 is rejected. There is sufficient evidence to indicate $\mu > 75$ at $\alpha = .10$.

d. For this two-tailed test, the p-value $= .1032$.

Since the p-value $= .1032 > \alpha = .01$, H_0 is not rejected. There is insufficient evidence to indicate $\mu \neq 75$ at $\alpha = .01$.

6.31 a. To test the manufacturer's claim, we test:

H_0: $\mu = 3.5$
H_a: $\mu < 3.5$

The test statistic is $z = \dfrac{\bar{x} - \mu_0}{\sigma_{\bar{x}}} = \dfrac{3.3 - 3.5}{\frac{66/60}{\sqrt{50}}} = -1.29$

The rejection region requires $\alpha = .05$ in the lower tail of the z-distribution. From Table III, Appendix A, $z_{.05} = 1.645$. The rejection region is $z < -1.645$.

Since the observed value of the test statistic does not fall in the rejection region ($z = -1.29 \not< -1.645$), H_0 is not rejected. There is insufficient evidence to support the manufacturer's claim at $\alpha = .05$.

b. p-value $= P(z \leq -1.29) = .5 - .4015 = .0985$

c. The smaller the p-value, the more support for the manufacturer's claim. The p-value measures the probability of observing your test statistic or anything more unusual if H_0 is true. As the p-value gets smaller, there is more evidence to reject H_0.

6.33 a. To determine if children in this age group perceive a risk associated with failure to wear helmets, we test:

H_0: $\mu = 2.5$
H_a: $\mu > 2.5$

b. The test statistic is $z = \dfrac{\bar{x} - \mu_0}{\sigma_{\bar{x}}} = \dfrac{3.39 - 2.5}{.80/\sqrt{797}} = 31.41$

p-value $= P(z \geq 31.41) \approx .5 - .5 = 0$

c. There is strong evidence to reject H_0 for any reasonable value of α. There is strong evidence to indicate the mean perceived risk associated with failure to wear helmets is greater than 2.5 for any reasonable value of α.

6.35 a. To test the designer's research hypothesis, we test:

H_0: $\mu = 250$
H_a: $\mu > 250$

b. The test statistic is $z = \dfrac{\bar{x} - \mu_0}{\sigma_{\bar{x}}} = \dfrac{256.3 - 250}{43.4/\sqrt{135}} = 1.69$

The p-value $= P(z \geq 1.69) = .5 - .4545 = .0455$

There is evidence to reject H_0 for $\alpha = .05$. There is evidence to indicate the mean driving distance now is greater than 250 yards at $\alpha = .05$.

c. The hypotheses would change for a two-tailed test. The new hypotheses would be:

H_0: $\mu = 250$
H_a: $\mu \neq 250$

The p-value for a two-tailed test would be two times the p-value for a one-tailed test. The p-value $= P(z \leq -1.69) + P(z \geq 1.69) = 2P(z \geq 1.69) = 2(.5 - .4545) = .0910$.

There is no evidence to reject H_0 for $\alpha = .05$. There is no evidence to indicate the mean driving distance now is different than 250 yards at $\alpha = .05$.

There is evidence to reject H_0 for $\alpha = .10$. There is evidence to indicate the mean driving distance now is different than 250 yards at $\alpha = .10$.

6.37 We should use the t-distribution in testing a hypothesis about a population mean if the sample size is small, the population being sampled from is normal, and the variance of the population is unknown.

6.39 a. $P(t > 1.440) = .10$
(using Table IV, Appendix A, with df $= 6$)

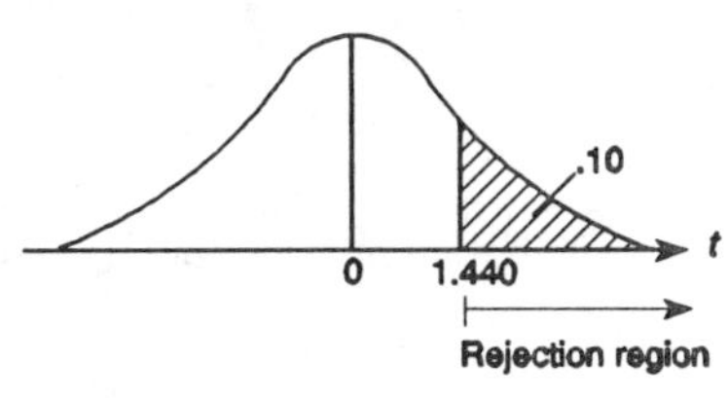

b. $P(t < -1.782) = .05$
(using Table IV, Appendix A, with df = 12)

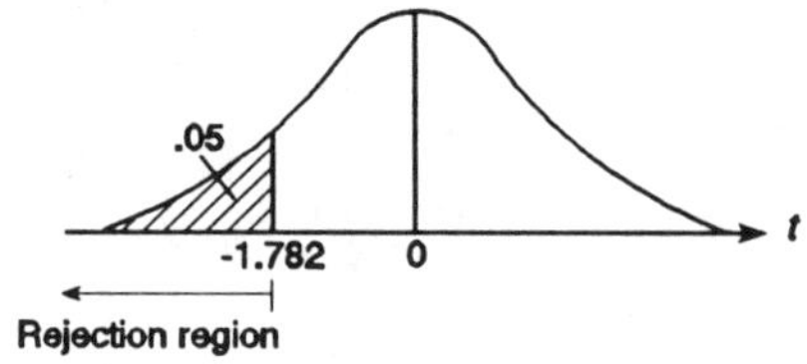

c. $P(t < -2.060) = P(t > 2.060) = .025$
(using Table IV, Appendix A, with df = 25)

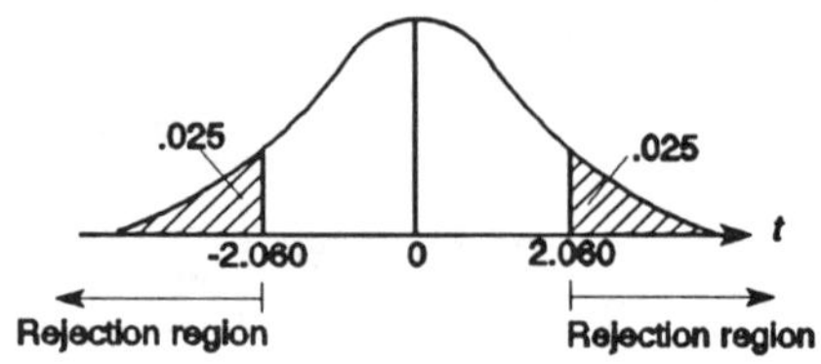

6.41 a. The rejection region requires $\alpha/2 = .05/2 = .025$ in each tail of the t-distribution with df $= n - 1 = 14 - 1 = 13$. From Table IV, Appendix A, $t_{.025} = 2.160$. The rejection region is $t < -2.160$ or $t > 2.160$.

b. The rejection region requires $\alpha = .01$ in the upper tail of the t-distribution with df $= n - 1 = 24 - 1 = 23$. From Table IV, Appendix A, $t_{.01} = 2.500$. The rejection region is $t > 2.500$.

c. The rejection region requires $\alpha = .10$ in the upper tail of the t-distribution with df $= n - 1 = 9 - 1 = 8$. From Table IV, Appendix A, $t_{.10} = 1.397$. The rejection region is $t > 1.397$.

d. The rejection region requires $\alpha = .01$ in the lower tail of the t-distribution with df $= n - 1 = 12 - 1 = 11$. From Table IV, Appendix A, $t_{.01} = 2.718$. The rejection region is $t < -2.718$.

e. The rejection region requires $\alpha/2 = .10/2 = .05$ in each tail of the t-distribution with df $= n - 1 = 20 - 1 = 19$. From Table IV, Appendix A, $t_{.05} = 1.729$. The rejection region is $t < -1.729$ or $t > 1.729$.

f. The rejection region requires $\alpha = .05$ in the lower tail of the t-distribution with df $= n - 1 = 4 - 1 = 3$. From Table IV, Appendix A, $t_{.05} = 2.353$. The rejection region is $t < -2.353$.

6.43 a. We must assume that a random sample was drawn from a normal population.

b. The hypotheses are:

H_0: $\mu = 1000$
H_a: $\mu > 1000$

The test statistic is $t = 1.894$.

The p-value is .0382.

There is evidence to reject H_0 for $\alpha > .0382$. There is evidence to indicate the mean is greater than 1000 for $\alpha > .0382$.

c. The hypotheses are:

H_0: $\mu = 1000$
H_a: $\mu \neq 1000$

The test statistic is $t = 1.894$.

The p-value is $2(.0382) = .0764$.

There is no evidence to reject H_0 for $\alpha = .05$. There is insufficient evidence to indicate the mean is different than 1000 for $\alpha = .05$.

There is evidence to reject H_0 for $\alpha > .0764$. There is evidence to indicate the mean is different than 1000 for $\alpha > .0764$.

6.45 a. From the printout, the alternative hypothesis is MU N.E. 1.000 or H_a: $\mu \neq 1.0$.

b. From the printout, the value of the test statistic is T = 1.44 or $t = 1.44$.

c. From the printout, the p-value is P VALUE 0.17 or p-value = 0.17. Since the p-value is not small, H_0 would not be rejected for any reasonable value of α. There is insufficient evidence to indicate the mean crop content of the spinifex pigeons is different from 1.0 for $\alpha < 0.17$.

6.47 a. Some preliminary calculations:

$$\bar{x} = \frac{\sum x}{n} = \frac{66}{3} = 22$$

$$s^2 = \frac{\sum x^2 - \frac{(\sum x)^2}{n}}{n - 1} = \frac{1460 - \frac{66^2}{3}}{3 - 1} = 4 \qquad s = \sqrt{4} = 2$$

To determine if the mean length of great white sharks off the Bermuda coast is in excess of 21 feet, we test:

H_0: $\mu = 21$
H_a: $\mu > 21$

The test statistic is $t = \dfrac{\bar{x} - \mu_0}{s/\sqrt{n}} = \dfrac{22 - 21}{2/\sqrt{3}} = 0.87$

The rejection region requires $\alpha = .10$ in the upper tail of the t-distribution with df $= n - 1 = 3 - 1 = 2$. From Table IV, Appendix A, $t_{.10} = 1.886$. The rejection region is $z > 1.886$.

Since the observed value of the test statistic does not fall in the rejection region ($t = 0.87 \not> 1.886$), H_0 is not rejected. There is insufficient evidence to indicate that the mean length of the great white sharks off the Bermuda coast is in excess of 21 feet at $\alpha = .10$.

b. p-value $= P(t \geq .87) > .10$ (from Table IV, Appendix A with df = 2)

c. We must be able to assume that the lengths of the great white sharks off the Bermuda coast are normally distributed and that a random sample has been obtained.

d. Those sharks which are slow/old/near shore/etc. are more likely to be selected, indicating a strong possibility that this is not a random sample.

6.49 To determine if the true mean PST score of all patients with mild to moderate traumatic brain injury exceeds the "normal" value of 40, we test:

H_0: $\mu = 40$
H_a: $\mu > 40$

The test statistic is $t = \dfrac{\bar{x} - \mu_0}{s/\sqrt{n}} = \dfrac{48.43 - 48}{20.76/\sqrt{23}} = 1.95$

The rejection region requires $\alpha = .05$ in the upper tail of the t-distribution with df $= n - 1 = 23 - 1 = 22$. From Table IV, Appendix A, $t_{.05} = 1.717$. The rejection region is $t > 1.717$.

Since the observed value of the test statistic falls in the rejection region ($t = 1.95 > 1.717$), H_0 is rejected. There is sufficient evidence to indicate that the true mean PST score of all patients with mild to moderate traumatic brain injury exceeds the "normal" value of 40 at $\alpha = .05$.

6.51 The sample size is large enough if the interval $p_0 \pm 3\sigma_{\hat{p}}$ is contained in the interval (0, 1).

a. $p_0 \pm 3\sqrt{\dfrac{p_0q_0}{n}} \Rightarrow .05 \pm 3\sqrt{\dfrac{(.05)(.95)}{500}} \Rightarrow .05 \pm .029 \Rightarrow (.021, .079)$

Since the interval is contained in the interval (0, 1), the sample size is large enough.

b. $p_0 \pm 3\sqrt{\dfrac{p_0q_0}{n}} \Rightarrow .99 \pm 3\sqrt{\dfrac{(.99)(.01)}{100}} \Rightarrow .99 \pm .030 \Rightarrow (.960, 1.020)$

Since the interval is not contained in the interval (0, 1), the sample size is not large enough.

c. $p_0 \pm 3\sqrt{\dfrac{p_0q_0}{n}} \Rightarrow .2 \pm 3\sqrt{\dfrac{(.2)(.8)}{50}} \Rightarrow .2 \pm .170 \Rightarrow (.030, .370)$

Since the interval is contained in the interval (0, 1), the sample size is large enough.

d. $p_0 \pm 3\sqrt{\dfrac{p_0q_0}{n}} \Rightarrow .2 \pm 3\sqrt{\dfrac{(.2)(.8)}{20}} \Rightarrow .2 \pm .268 \Rightarrow (-.068, 468)$

Since the interval is not contained in the interval (0, 1), the sample size is not large enough.

e. $p_0 \pm 3\sqrt{\dfrac{p_0q_0}{n}} \Rightarrow .4 \pm 3\sqrt{\dfrac{(.4)(.6)}{10}} \Rightarrow .4 \pm .465 \Rightarrow (-.065, .865)$

Since the interval is not contained in the interval (0, 1), the sample size is not large enough.

6.53 a. $z = \dfrac{\hat{p} - p_0}{\sqrt{\dfrac{p_0q_0}{n}}} = \dfrac{.84 - .9}{\sqrt{\dfrac{.9(.1)}{100}}} = -2.00$

b. The denominator in Exercise 6.52 is $\sqrt{\frac{.75(.25)}{100}} = .0433$ as compared to $\sqrt{\frac{.9(.1)}{100}} = .03$ in part **a**. Since the denominator in this problem is smaller, the absolute value of z is larger.

c. The rejection region requires $\alpha = .05$ in the lower tail of the z-distribution. From Table III, Appendix A, $z_{.05} = 1.645$. The rejection region is $z < -1.645$.

Since the observed value of the test statistic falls in the rejection region ($z = -2.00 < -1.645$), H_0 is rejected. There is sufficient evidence to indicate the population proportion is less than .9 at $\alpha = .05$.

d. The p-value $= P(z \leq -2.00) = .5 - .4772 = .0228$ (from Table III, Appendix A).

6.55 a. No. The p-value is the probability of observing your test statistic or anything more unusual if H_0 is true. For this problem, the p-value $= P(z \geq .44) = .3300$.

b. If the alternative hypothesis were two-tailed, the p-value would be 2 times the p-value for a one-tailed test. For this problem, the p-value $= 2(.3300) = .6600$.

There is no evidence to reject H_0 for $\alpha \leq .10$. There is no evidence to indicate that $p > .5$ for $\alpha \leq .10$.

6.57 First, check to see if the normal approximation is adequate:

$$p_0 \pm 3\sigma_{\hat{p}} \Rightarrow p_0 \pm 3\sqrt{\frac{p_0 q_0}{n}} \Rightarrow .5 \pm 3\sqrt{\frac{(.5)(.5)}{7000}} \Rightarrow .5 \pm .018 \Rightarrow (.482, .518)$$

Since the interval falls completely in the interval (0, 1), the normal distribution will be adequate.

To determine if there was a placebo effect, we test:

H_0: $p = .5$
H_a: $p > .5$

The test statistic is $z = \dfrac{\hat{p} - p_0}{\sqrt{\dfrac{p_0 q_0}{n}}} = \dfrac{.7 - .5}{\sqrt{\dfrac{(.5)(.5)}{7000}}} = 33.47$

The rejection region requires $\alpha = .05$ in the upper tail of the z-distribution. From Table III, Appendix A, $z_{.05} = 1.645$. The rejection region is $z > 1.645$.

Since the observed value of the test statistic falls in the rejection region ($z = 33.47 > 1.645$), H_0 is rejected. There is sufficient evidence to indicate that the true proportion of patients who improved (while on the placebo) is greater than .5 at $\alpha = .05$.

6.59 a. First, check to see if the normal approximation is adequate:

$$p_0 \pm 3\sigma_{\hat{p}} \Rightarrow p_0 \pm 3\sqrt{\frac{p_0 q_0}{n}} \Rightarrow .25 \pm 3\sqrt{\frac{(.25)(.75)}{159}} \Rightarrow .25 \pm .103 \Rightarrow (.147, .353)$$

Since the interval falls completely in the interval (0, 1), the normal distribution will be adequate.

$$\hat{p} = \frac{x}{n} = \frac{124}{159} = .786$$

To determine if the percentage of truckers who suffer from sleep apnea differs from 25%, we test:

H_0: $p = .25$
H_a: $p \neq .25$

The test statistic is $z = \dfrac{\hat{p} - p_0}{\sqrt{\dfrac{p_0 q_0}{n}}} = \dfrac{.786 - .25}{\sqrt{\dfrac{(.25)(.75)}{159}}} = 15.61$

The rejection region requires $\alpha/2 = .10/2 = .05$ in each tail of the z-distribution. From Table III, Appendix A, $z_{.05} = 1.645$. The rejection region is $z < -1.645$ or $z > 1.645$.

Since the observed value of the test statistic falls in the rejection region ($z = 15.61 > 1.645$), H_0 is rejected. There is sufficient evidence to indicate that the percentage of truckers who suffer from sleep apnea differs from 25% at $\alpha = .05$.

b. The observed significance level is the p-value and is:

$$p\text{-value} = P(z \geq 15.61) + P(z \leq -15.61) \approx (.5 - .5) + (.5 - .5) = 0$$

Since the p-value is so small, we would reject H_0 for any reasonable value of α. There is sufficient evidence to indicate that the percentage of truckers who suffer from sleep apnea differs from 25%.

c. The inference from a confidence interval and a test of hypothesis must agree because the same numbers are used in both if the same level of significance is used.

6.61 First, check to see if the normal approximation is adequate:

$$p_0 \pm 3\sigma_{\hat{p}} \Rightarrow p_0 \pm 3\sqrt{\frac{p_0 q_0}{n}} \Rightarrow .5 \pm 3\sqrt{\frac{(.5)(.5)}{2,237}} \Rightarrow .5 \pm .032 \Rightarrow (.468, .532)$$

Since the interval falls completely in the interval (0, 1), the normal distribution will be adequate.

$$\hat{p} = \frac{x}{n} = \frac{1,630}{2,237} = .729$$

To determine if the true percentage of engineering Ph.D. degrees awarded to foreign nationals exceeds 50%, we test:

H_0: $p = .5$
H_a: $p > .5$

The test statistic is $z = \dfrac{\hat{p} - p_0}{\sqrt{\dfrac{p_0 q_0}{n}}} = \dfrac{.729 - .5}{\sqrt{\dfrac{(.5)(.5)}{2,237}}} = 21.66$

The rejection region requires $\alpha = .01$ in the upper tail of the z-distribution. From Table III, Appendix A, $z_{.01} = 2.33$. The rejection region is $z > 2.33$.

Since the observed value of the test statistic falls in the rejection region ($z = 21.66 > 2.33$), H_0 is rejected. There is sufficient evidence to indicate that the true percentage of engineering Ph.D. degrees awarded to foreign nationals exceeds 50% at $\alpha = .01$.

6.63 a. Since the normal distribution is symmetric, the probability that a randomly selected observation exceeds the mean of a normal distribution is .5.

b. By the definition of "median," the probability that a randomly selected observation exceeds the median of a normal distribution is .5.

c. If the distribution is not normal, the probability that a randomly selected observation exceeds the mean depends on the distribution. With the information given, the probability cannot be determined.

d. By definition of "median," the probability that a randomly selected observation exceeds the median of a non-normal distribution is .5.

6.65 a. H_0: $\eta = 9$
H_a: $\eta > 9$

The test statistic is $S = \{\text{Number of observations greater than } 9\} = 7$.

The p-value $= P(x \geq 7)$ where x is a binomial random variable with $n = 10$ and $p = .5$. From Table II,

$$p\text{-value} = P(x \geq 7) = 1 - P(x \leq 6) = 1 - .828 = .172$$

Since the p-value $= .172 > \alpha = .05$, H_0 is not rejected. There is insufficient evidence to indicate the median is greater than 9 at $\alpha = .05$.

b. H_0: $\eta = 9$
H_a: $\eta \neq 9$

$S_1 = \{\text{Number of observations less than } 9\} = 3$ and
$S_2 = \{\text{Number of observations greater than } 9\} = 7$

The test statistic is S = larger of S_1 and $S_2 = 7$.

The p-value $= 2P(x \geq 7)$ where x is a binomial random variable with $n = 10$ and $p = .5$. From Table II,

$$p\text{-value} = 2P(x \geq 7) = 2(1 - P(x \leq 6)) = 2(1 - .828) = .344$$

Since the p-value $= .344 > \alpha = .05$, H_0 is not rejected. There is insufficient evidence to indicate the median is different than 9 at $\alpha = .05$.

c. H_0: $\eta = 20$
H_a: $\eta < 20$

The test statistic is $S = \{\text{Number of observations less than } 20\} = 9$

The p-value $= P(x \geq 9)$ where x is a binomial random variable with $n = 10$ and $p = .5$. From Table II,

$$p\text{-value} = P(x \geq 9) = 1 - P(x \leq 8) = 1 - .989 = .011$$

Since the p-value $= .011 < \alpha = .05$, H_0 is rejected. There is sufficient evidence to indicate the median is less than 20 at $\alpha = .05$.

d. $H_0\text{: } \eta = 20$
$H_a\text{: } \eta \neq 20$

$S_1 = \{\text{Number of observations less than 20}\} = 9$ and
$S_2 = \{\text{Number of observations greater than 20}\} = 1$

The test statistic is S = larger of S_1 and $S_2 = 9$.

The p-value $= 2P(x \geq 9)$ where x is a binomial random variable with $n = 10$ and $p = .5$. From Table II,

$$p\text{-value} = 2P(x \geq 9) = 2(1 - P(x \leq 8)) = 2(1 - .989) = .022$$

Since the p-value $= .022 < \alpha = .05$, H_0 is rejected. There is sufficient evidence to indicate the median is different than 20 at $\alpha = .05$.

e. For all parts, $\mu = np = 10(.5) = 5$ and $\sigma = \sqrt{npq} = \sqrt{10(.5)(.5)} = 1.581$.

For part **a**, $P(x \geq 7) \approx P\left[z \geq \dfrac{(7 - .5) - 5}{1.581}\right] = P(z \geq .95) = .5 - .3289 = .1911$

This is close to the probability .172 in part **a**. The conclusion is the same.

For part **b**, $2P(x \geq 7) \approx 2P\left[z \geq \dfrac{(7 - .5) - 5}{1.581}\right] = 2P(z \geq .95) = 2(.5 - .3289)$
$= .3422$

This is close to the probability .344 in part **b**. The conclusion is the same.

For part **c**, $P(x \geq 9) \approx P\left[z \geq \dfrac{(9 - .5) - 5}{1.581}\right] = P(z \geq 2.21) = .5 - .4864$
$= .0136$

This is close to the probability .011 in part **c**. The conclusion is the same.

For part **d**, $2P(x \geq 9) \approx 2P\left[z \geq \dfrac{(9 - .5) - 5}{1.581}\right] = 2P(z \geq 2.21) = 2(.5 - .4864)$
$= .0272$

This is close to the probability .022 in part **d**. The conclusion is the same.

f. We must assume only that the sample is selected randomly from a continuous probability distribution.

6.67 To determine whether the median is less than 10, we test:

$H_0\text{: } \eta = 20$
$H_a\text{: } \eta < 20$

The test statistic is $z = \dfrac{(S - .5) - .5n}{.5\sqrt{n}} = \dfrac{(25 - .5) - .5(30)}{.5\sqrt{30}} = 3.47$
(where S = number of observations less than 20)

The rejection region requires $\alpha = .05$ in the upper tail of the z-distribution. From Table III, Appendix A, $z_{.05} = 1.645$. The rejection region is $z > 1.645$.

Since the observed value of the test statistic falls in the rejection region ($z = 3.47 > 1.645$), H_0 is rejected. There is sufficient evidence to indicate that the median number of fish remaining in the left compartment is less than 20 at $\alpha = .05$.

6.69 To determine if the median remission time is increased by the new treatment, we test:

$$H_0\text{: } \eta = 4.5$$
$$H_a\text{: } \eta > 4.5$$

The test statistic is $S = \{\text{Number of measurements greater than } 4.5\} = 6$.

The p-value $= P(x \geq 6)$ where x is a binomial random variable with $n = 7$ and $p = .5$. From Table II, Appendix A,

$$p\text{-value} = P(x \geq 6) = 1 - P(x \leq 5) = 1 - .937 = .063$$

If $\alpha = .05$, H_0 is not rejected because p-value $= .063 > \alpha = .05$. There is insufficient evidence to indicate the median remission time is increased by the new treatment.

If $\alpha = .10$, H_0 is rejected because p-value $= .063 < \alpha = .10$. There is sufficient evidence to indicate the median remission time is increased by the new treatment.

6.71 To determine if the median lymphocyte count of all West Indian/African workers exceeds 20, we test:

$$H_0\text{: } \eta = 20$$
$$H_a\text{: } \eta > 20$$

The test statistic is $S = 25$ (where S = number of observations greater than 20 from the printout).

The observed p-value of the test is $p = .2257$. Since the p-value is greater than α ($p = .2257 > .05$), H_0 is not rejected. There is insufficient evidence to indicate that the median lymphocyte count of all West Indian/African workers is greater than 20 at $\alpha = .05$.

6.73 For a large sample test of hypothesis about a population mean, no assumptions are necessary because the Central Limit Theorem assures that the test statistic will be approximately normally distributed. For a small sample test of hypothesis about a population mean, we must assume that the population being sampled from is normal. The test statistic for the large sample test is the z statistic, and the test statistic for the small sample test is the t statistic.

6.75 The larger the p-value associated with a test of hypothesis, the stronger the support for the **null** hypothesis. The p-value is the probability of observing your test statistic or anything more unusual, given the null hypothesis is true. If this value is large, it would not be very unusual to observe this test statistic if the null hypothesis were true. Thus, it would lend support that the null hypothesis is true.

6.77 a. $H_a\text{: } p = .35$
$H_a\text{: } p < .35$

The test statistic is $z = \dfrac{\hat{p} - p_0}{\sqrt{\dfrac{p_0 q_0}{n}}} = \dfrac{.29 - .35}{\sqrt{\dfrac{.35(.65)}{200}}} = -1.78$

The rejection region requires $\alpha = .05$ in the lower tail of the z-distribution. From Table III, Appendix A, $z_{.05} = 1.645$. The rejection region is $z < -1.645$.

Since the observed value of the test statistic falls in the rejection region ($z = -1.78 < -1.645$), H_0 is rejected. There is sufficient evidence to indicate $p < .35$ at $\alpha = .05$.

b. H_0: $p = .35$
H_a: $p \neq .35$

The test statistic is $z = -1.78$ (from a).

The rejection region requires $\alpha/2 = .05/2 = .025$ in each tail of the z-distribution. From Table III, Appendix A, $t_{.025} = 1.96$. The rejection region is $z < -1.96$ or $z > 1.96$.

Since the observed value of the test statistic does not fall in the rejection region ($z = -1.78 \not< -1.96$), H_0 is not rejected. There is insufficient evidence to indicate p is different from .35 at $\alpha = .05$.

6.79 a. The p-value $= .1288 = P(t \geq 1.174)$. Since the p-value is not very small, there is no evidence to reject H_0 for $\alpha \leq .10$. There is no evidence to indicate the mean is greater than 10.

b. We must assume that the sampled population is normally distributed.

c. For the alternative hypothesis H_a: $\mu \neq 10$, the p-value is 2 times the p-value for the one-tailed test. The p-value $= 2(.1288) = .2576$. There is no evidence to reject H_0 for $\alpha \leq .10$. There is no evidence to indicate the mean is different from 10.

6.81 a. To determine if the true mean score of all sleep-deprived subjects is less than 80, we test:

H_0: $\mu = 80$
H_a: $\mu < 80$

The test statistic is $t = \dfrac{\bar{x} - \mu_0}{s/\sqrt{n}} = \dfrac{63 - 80}{17/\sqrt{12}} = -3.46$

The rejection region requires $\alpha = .05$ in the lower tail of the t-distribution with df $= n - 1 = 12 - 1 = 11$. From Table IV, Appendix A, $t_{.05} = 1.796$. The rejection region is $t < 1.796$.

Since the observed value of the test statistic falls in the rejection region ($t = -3.46 < -1.796$), H_0 is rejected. There is sufficient evidence to indicate that the true mean score of all sleep-deprived subjects is less than 80 at $\alpha = .05$.

b. We must assume that the overall test scores of sleep-deprived students are normally distributed.

6.83 a. The test statistic is $t = \dfrac{\bar{x} - \mu_0}{s/\sqrt{n}} = \dfrac{1173.6 - 1100}{36.3/\sqrt{3}} = 3.512$

The p-value $= P(t \geq 3.512)$. From Table IV with df $= n - 1 = 3 - 1 = 2$, $.025 < p\text{-value} < .05$.

b. The p-value $= .0362 = P(t \geq 3.512)$. Since this p-value is fairly small, there is evidence to reject H_0 for $\alpha > .0362$. There is evidence to indicate the mean length of life of a certain mechanical component is longer than 1100 hours.

c. We must assume the population of lifetimes is normally distributed.

d. A Type I error would be of most concern for this test. A Type I error would be concluding the mean lifetime is greater than 1100 hours when in fact the mean lifetime is not greater than 1100.

6.85 a. First, check to see if the normal approximation is adequate:

$$p_0 \pm 3\sigma_{\hat{p}} \Rightarrow p_0 \pm 3\sqrt{\frac{p_0 q_0}{n}} \Rightarrow .75 \pm 3\sqrt{\frac{(.75)(.25)}{470}} \Rightarrow .75 \pm .060 \Rightarrow (.690, .810)$$

Since the interval falls completely in the interval (0, 1), the normal distribution will be adequate.

Let p = proportion of golf balls sold that are white. $\hat{p} = \dfrac{410}{470} = .872$

To determine if the manufacturer's claim is true, we test:

H_0: $p = .75$
H_a: $p > .75$

The test statistic is $z = \dfrac{\hat{p} - p_0}{\sqrt{\dfrac{p_0 q_0}{n}}} = \dfrac{.872 - .75}{\sqrt{\dfrac{.75(.25)}{470}}} = 6.11$

The rejection region requires $\alpha = .01$ in the upper tail of the z-distribution. From Table III, Appendix A, $z_{.01} = 2.33$. The rejection region is $z > 2.33$.

Since the observed value of the test statistic falls in the rejection region ($z = 6.11 > 2.33$), H_0 is rejected. There is sufficient evidence to indicate more than 75% of all golf balls sold are white at $\alpha = .01$.

b. The p-value $= P(z \geq 6.11) \approx .5 - .5 = 0$. There is evidence to reject H_0 for $\alpha > .0001$.

6.87 a. To determine if the production process should be halted, we test:

H_0: $\mu = 3$
H_a: $\mu > 3$

where μ = mean amount of PCB in the effluent.

The test statistic is $z = \dfrac{\bar{x} - \mu_0}{\sigma_{\bar{x}}} = \dfrac{3.1 - 3}{.5/\sqrt{50}} = 1.41$

The rejection region requires $\alpha = .01$ in the upper tail of the z-distribution. From Table III, Appendix A, $z_{.01} = 2.33$. The rejection region is $z > 2.33$.

Since the observed value of the test statistic does not fall in the rejection region, ($z = 1.41 \ngtr 2.33$), H_0 is not rejected. There is insufficient evidence to indicate the mean amount of PCB in the effluent is more than 3 parts per million at $\alpha = .01$. Do not halt the manufacturing process.

b. As plant manager, I do not want to shut down the plant unnecessarily. Therefore, I want $\alpha = P(\text{shut down plant when } \mu = 3)$ to be small.

6.89 a. First, check to see if the normal approximation will be adequate:

$$p_0 \pm 3\sigma_{\hat{p}} \Rightarrow p_0 \pm 3\sqrt{\frac{p_0 q_0}{n}} \Rightarrow .02 \pm 3\sqrt{\frac{.02(.98)}{400}} \Rightarrow .02 \pm .021 \Rightarrow (-.001, .041)$$

Since the interval does not lie completely in the interval (0, 1), the normal approximation will not be adequate.

We will use the approximation to get an idea of the answer.

$$\hat{p} = \frac{x}{n} = \frac{11}{400} = .0275$$

To determine if the proportion of defective can openers in the shipment exceeds .02, we test:

H_0: $p = .02$
H_a: $p > .02$

The test statistic is $z = \dfrac{\hat{p} - p_0}{\sigma_{\hat{p}}} = \dfrac{\hat{p} - p_0}{\sqrt{\dfrac{p_0 q_0}{n}}} = \dfrac{.0275 - .02}{\sqrt{\dfrac{.02(.98)}{400}}} = 1.07$

The rejection region requires $\alpha = .05$ in the upper tail of the z-distribution. From Table IV, Appendix A, $z_{.05} = 1.645$. The rejection region is $z > 1.645$.

Since the observed value of the test statistic does not fall in the rejection region ($z = 1.07 \ngtr 1.645$), H_0 is not rejected. There is insufficient evidence to indicate the proportion of defective can openers exceeds .02 at $\alpha = .05$.

b. From part a, $z = \dfrac{\hat{p} - p_0}{\sqrt{\dfrac{p_0 q_0}{n}}} = \dfrac{.0275 - .02}{\sqrt{\dfrac{.02(.98)}{400}}} = 1.07$

The p-value $= P(z \geq 1.07) = .5 - .3577 = .1423$ from Table III, Appendix A.

Since the p-value is not small, there is no evidence to reject H_0. There is no evidence to indicate the proportion of defectives exceeds .02.

6.91 a. To determine if the average high technology stock is riskier than the market as a whole, we test:

H_0: $\mu = 1$
H_a: $\mu > 1$

b. The test statistic is $t = \dfrac{\bar{x} - \mu_0}{s/\sqrt{n}}$

The rejection region requires $\alpha = .10$ in the upper tail of the t-distribution with df $= n - 1 = 15 - 1 = 14$. From Table IV, Appendix A, $t_{.10} = 1.345$. The rejection region is $t > 1.345$.

c. We must assume the population of beta coefficients of technology stocks is normally distributed.

d. The test statistic is $t = \dfrac{\bar{x} - \mu_0}{s/\sqrt{n}} = \dfrac{1.23 - 1}{.37/\sqrt{15}} = 2.41$

Since the observed value of the test statistic falls in the rejection region ($t = 2.41 > 1.345$), H_0 is rejected. There is sufficient evidence to indicate the mean high technology stock is riskier than the market as a whole at $\alpha = .10$.

e. From Table IV, Appendix A, with df $= n - 1 = 15 - 1 = 14$, $.01 < P(t \geq 2.41) < .025$. Thus, $.01 < p$-value $< .025$. The probability of observing this test statistic, $t = 2.41$, or anything more unusual is between .01 and .025. Since this probability is small, there is evidence to indicate the null hypothesis is false for $\alpha = .05$.

6.93 First, check to see if the normal approximation is adequate:

$$p_0 \pm 3\sigma_{\hat{p}} \Rightarrow p_0 \pm 3\sqrt{\frac{p_0 q_0}{n}} \Rightarrow .85 \pm 3\sqrt{\frac{.85(.15)}{.60}} \Rightarrow .85 \pm .138 \Rightarrow (.712, .988)$$

Since the interval does not include 0 or 1, the normal approximation is appropriate.

$\hat{p} = \dfrac{47}{60} = .783$. To determine if the data contradict the claim, we test:

H_0: $p = .85$
H_a: $p < .85$

The test statistic is $z = \dfrac{\hat{p} - p_0}{\sqrt{\dfrac{p_0 q_0}{n}}} = \dfrac{.783 - .85}{\sqrt{\dfrac{.85(.15)}{60}}} = -1.45$

The rejection region requires $\alpha = .05$ in the lower tail of the z-distribution. From Table III, Appendix A, $z_{.05} = 1.645$. The rejection region is $z < -1.645$.

Since the observed value of the test statistic does not fall in the rejection region ($z = -1.45 \not< -1.645$), H_0 is not rejected. There is insufficient evidence to indicate the claim is incorrect at $\alpha = .05$.

6.95 To determine whether the median GPA exceeds 3.0, we test:

H_0: $\eta = 3$
H_a: $\eta > 3$

The test statistic is $S = 5$ (where S = number of observations greater than 3)

The p-value $= P(x \geq 5)$ where x is a binomial random variable with $n = 8$ and $p = .5$. From Table II,

$$p\text{-value} = P(x \geq 5) = 1 - P(x \leq 4) = 1 - .637 = .363.$$

Since the p-value $= .363 > \alpha = .05$, H_0 is not rejected. There is insufficient evidence to indicate that the median GPA exceeds 3.0 at $\alpha = .05$.

The observed significance level is p-value $= .363$. If the median GPA is 3.0, then the probability of observing 5 or more GPA's greater than or equal to 3.0 out of 8 is .363. This is not an unusual probability. There is no indication that the median GPA is greater than 3.0.

Chapter 7

Comparing Population Means

7.1 a. For confidence coefficient .95, $\alpha = .05$ and $\alpha/2 = .025$. From Table III, Appendix A, $z_{.025} = 1.96$. The confidence interval is:

$$(\bar{x}_1 - \bar{x}_2) \pm z_{.025}\sqrt{\frac{\sigma_1^2}{n_1} + \frac{\sigma_2^2}{n_2}} \Rightarrow (5{,}275 - 5{,}240) \pm 1.96\sqrt{\frac{150^2}{400} + \frac{200^2}{400}}$$

$$\Rightarrow 35 \pm 24.5 \Rightarrow (10.5, 59.5)$$

We are 95% confident that the difference between the population means is between 10.5 and 59.5.

b. The test statistic is $z = \dfrac{(\bar{x}_1 - \bar{x}_2) - (\mu_1 - \mu_2)}{\sqrt{\dfrac{\sigma_1^2}{n_1} + \dfrac{\sigma_2^2}{n_2}}} = \dfrac{(5275 - 5240) - 0}{\sqrt{\dfrac{150^2}{400} + \dfrac{200^2}{400}}} = 2.8$

The p-value of the test is $P(z \leq -2.8) + P(z \geq 2.8) = 2P(z \geq 2.8) = 2(.5 - .4974)$
$= 2(.0026) = .0052$

Since the p-value is so small, there is evidence to reject H_0. There is evidence to indicate the two population means are different for $\alpha > .0052$.

c. The p-value would be half of the p-value in part **b**. The p-value $= P(z \geq 2.8) = .5 - .4974 = .0026$. Since the p-value is so small, there is evidence to reject H_0. There is evidence to indicate the mean for population 1 is larger than the mean for population 2 for $\alpha > .0026$.

d. The test statistic is $z = \dfrac{(\bar{x}_1 - \bar{x}_2) - (\mu_1 - \mu_2)}{\sqrt{\dfrac{\sigma_1^2}{n_1} + \dfrac{\sigma_2^2}{n_2}}} = \dfrac{(5275 - 5240) - 25}{\sqrt{\dfrac{150^2}{400} + \dfrac{200^2}{400}}} = .8$

The p-value of the test is $P(z \leq -.8) + P(z \geq .8) = 2P(z \geq .8) = 2(.5 - .2881)$
$= 2(.2119) = .4238$

Since the p-value is so large, there is no evidence to reject H_0. There is no evidence to indicate that the difference in the 2 population means is different from 25 for $\alpha \leq .10$.

e. We must assume that we have two independent random samples.

7.3 a. $\mu_{\bar{x}_1} = \mu_1 = 14$

$$\sigma_{\bar{x}_1} = \frac{\sigma_1}{\sqrt{n_1}} = \frac{4}{\sqrt{100}} = .4$$

b. $\mu_{\bar{x}_2} = \mu_2 = 10$

$$\sigma_{\bar{x}_2} = \frac{\sigma_2}{\sqrt{n_2}} = \frac{3}{\sqrt{100}} = .3$$

c. $\mu_{\bar{x}_1-\bar{x}_2} = \mu_1 - \mu_2 = 14 - 10 = 4$

$$\sigma_{\bar{x}_1-\bar{x}_2} = \sqrt{\frac{\sigma_1^2}{n_1} + \frac{\sigma_2^2}{n_2}} = \sqrt{\frac{4^2}{100} + \frac{3^2}{100}} = \sqrt{\frac{25}{100}} = .5$$

d. Since $n_1 \geq 30$ and $n_2 \geq 30$, the sampling distribution of $\bar{x}_1 - \bar{x}_2$ is approximately normal by the Central Limit Theorem.

7.5 a. No. Both populations must be normal.

b. No. Both populations variances must be equal.

c. No. Both populations must be normal.

d. Yes.

e. No. Both populations must be normal.

7.7 a. $$s_p^2 = \frac{(n_1 - 1)s_1^2 + (n_2 - 1)s_2^2}{n_1 + n_2 - 2} = \frac{(25 - 1)200 + (25 - 1)180}{25 + 25 - 2} = \frac{9120}{48} = 190$$

b. $$s_p^2 = \frac{(20 - 1)25 + (10 - 1)40}{20 + 10 - 2} = \frac{835}{28} = 29.8214$$

c. $$s_p^2 = \frac{(8 - 1).20 + (12 - 1).30}{8 + 12 - 2} = \frac{4.7}{18} = .2611$$

d. $$s_p^2 = \frac{(16 - 1)2500 + (17 - 1)1800}{16 + 17 - 2} = \frac{66,300}{31} = 2138.7097$$

e. s_p^2 falls nearer the variance with the larger sample size.

7.9 a. $$\sigma_{\bar{x}_1-\bar{x}_2} = \sqrt{\frac{\sigma_1^2}{n_1} + \frac{\sigma_2^2}{n_2}} = \sqrt{\frac{100}{100} + \frac{64}{100}} = \sqrt{1.64} = 1.2806$$

b. The sampling distribution of $\bar{x}_1 - \bar{x}_2$ is approximately normal by the Central Limit Theorem since $n_1 \geq 30$ and $n_2 \geq 30$.

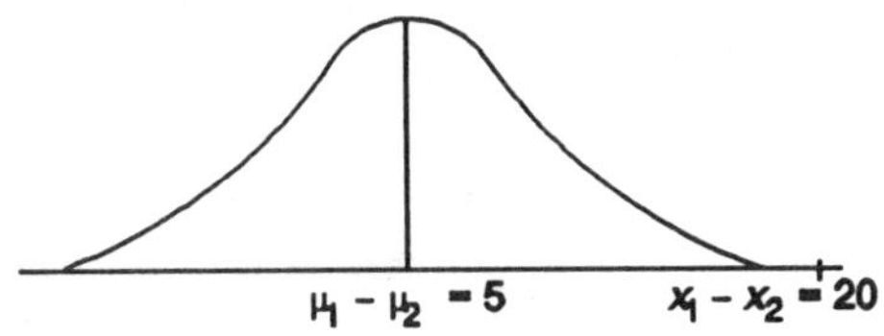

c. $\bar{x}_1 - \bar{x}_2 = 70 - 50 = 20$

Yes, it appears that $\bar{x}_1 - \bar{x}_2 = 20$ contradicts the null hypothesis H_0: $\mu_1 - \mu_2 = 5$.

d. The rejection region requires $\alpha/2 = .025 = .05/2$ in each tail of the z-distribution. From Table III, Appendix A, $z_{.025} = 1.96$. The rejection region is $z < -1.96$ or $z > 1.96$.

e. H_0: $\mu_1 - \mu_2 = 5$
H_a: $\mu_1 - \mu_2 \neq 5$

The test statistic is $$z = \frac{(\bar{x}_1 - \bar{x}_2) - 5}{\sqrt{\dfrac{\sigma_1^2}{n_1} + \dfrac{\sigma_2^2}{n_2}}} = \frac{(70 - 50) - 5}{1.2806} = 11.71$$

The rejection region is $z < -1.96$ or $z > 1.96$. (Refer to part **d.**)

Since the observed value of the test statistic falls in the rejection region ($z = 11.71 > 1.96$), H_0 is rejected. There is sufficient evidence to indicate the difference in the population means is not equal to 5 at $\alpha = .05$.

7.11 a. The test statistic is $z = -1.576$ and the p-value $= .1150$.

Since the p-value is not small, there is no evidence to reject H_0. There is no evidence to indicate that the population means are different for $\alpha \leq .10$.

b. The p-value would be half of the p-value in part a. The p-value $= 1/2(.1150) = .0575$.

There is no evidence to reject H_0 for $\alpha = .05$. There is no evidence to indicate that the mean for population 1 is less than the mean for population 2 for $\alpha = .05$.

There is evidence to reject H_0 for $\alpha > .0575$. There is evidence to indicate that the mean for population 1 is less than the mean for population 2 for $\alpha > .0575$.

7.13 a. $$s_p^2 = \frac{(n_1 - 1)s_1^2 + (n_2 - 1)s_2^2}{n_1 + n_2 - 2} = \frac{(17 - 1)3.4^2 + (12 - 1)4.8^2}{17 + 12 - 2} = 16.237$$

The test statistic is:

$$t = \frac{(\bar{x}_1 - \bar{x}_2) - 0}{\sqrt{s_p^2\left(\dfrac{1}{n_1} + \dfrac{1}{n_2}\right)}} = \frac{(5.4 - 7.9) - 0}{\sqrt{16.237\left(\dfrac{1}{17} + \dfrac{1}{12}\right)}} = -1.646$$

The p-value $= P(t \leq -1.646) + P(t \geq 1.646) = 2P(t \geq 1.646)$.

Using Table IV with df = $n_1 + n_2 = 17 + 12 - 2 = 27$, $P(t \geq 1.646)$ is between .05 and .10. Thus, $2(.05) < p\text{-value} < 2(.10)$ or $.10 < p\text{-value} < .20$.

Since the p-value is not small, there is no evidence to reject H_0. There is no evidence to indicate the means are different for $\alpha \leq .10$.

b. For confidence coefficient .95, $\alpha = .05$ and $\alpha/2 = .025$. From Table IV, Appendix A, with df $= n_1 + n_2 - 2 = 17 + 12 - 2 = 27$, $t_{.025} = 2.052$. The confidence interval is:

$$(\bar{x}_1 - \bar{x}_2) \pm t_{.025}\sqrt{s_p^2\left(\frac{1}{n_1} + \frac{1}{n_2}\right)} \quad \text{where } t \text{ has 27 df}$$

$$\Rightarrow (5.4 - 7.9) \pm 2.052\sqrt{16.237\left(\frac{1}{17} + \frac{1}{12}\right)} \Rightarrow -2.50 \pm 3.12 \Rightarrow (-5.62, 0.62)$$

7.15 a. Let μ_1 = mean degree of swelling for mice treated with bear bile and μ_2 = mean degree of swelling for mice treated with pig bile.

$$s_p^2 = \frac{(n_1 - 1)s_1^2 + (n_2 - 1)s_2^2}{n_1 + n_2 - 2} = \frac{(10 - 1)4.17^2 + (10 - 1)3.33^2}{10 + 10 - 2} = 14.2389$$

For confidence coefficient .95, $\alpha = .05$ and $\alpha/2 = .05/2 = .025$. From Table IV, Appendix A, with df $= n_1 + n_2 - 2 = 10 + 10 - 2 = 18$, $t_{.025} = 2.101$. The confidence interval is:

$$(\bar{x}_1 - \bar{x}_2) \pm t_{.025}\sqrt{s_p^2\left(\frac{1}{n_1} + \frac{1}{n_2}\right)} \Rightarrow (9.19 - 9.71) \pm 2.101\sqrt{14.289\left(\frac{1}{10} + \frac{1}{10}\right)}$$
$$\Rightarrow -.52 \pm 3.546 \Rightarrow (-4.066, 3.026)$$

We are 95% confident that the difference in the mean degree of swelling for mice treated with bear bile and mice treated with pig bile is between -4.066 and 3.026.

b. We must assume that the distribution of the degree of swelling for all mice receiving bear bile and the distribution of the degree of swelling for all mice receiving pig bile are normal. Also, we must assume that the variances of the two populations are equal.

c. Let μ_3 = mean degree of swelling for mice treated with saline and μ_2 = mean degree of swelling for mice treated with pig bile.

$$s_p^2 = \frac{(n_3 - 1)s_3^2 + (n_2 - 1)s_2^2}{n_3 + n_2 - 2} = \frac{(10 - 1)3.38^2 + (10 - 1)3.33^2}{10 + 10 - 2} = 11.25665$$

For confidence coefficient .95, $\alpha = .05$ and $\alpha/2 = .05/2 = .025$. From Table IV, Appendix A, with df $= n_1 + n_2 - 2 = 10 + 10 - 2 = 18$, $t_{.025} = 2.101$. The confidence interval is:

$$(\bar{x}_3 - \bar{x}_2) \pm t_{.025}\sqrt{s_p^2\left(\frac{1}{n_3} + \frac{1}{n_2}\right)} \Rightarrow (18.30 - 9.71) \pm 2.101\sqrt{11.25665\left(\frac{1}{10} + \frac{1}{10}\right)}$$
$$\Rightarrow 8.59 \pm 3.152 \Rightarrow (5.438, 11.742)$$

We are 95% confident that the difference in the mean degree of swelling for mice treated with saline and mice treated with pig bile is between 5.438 and 11.742.

7.17 Let μ_1 = mean number of timeouts for students in Option III classrooms and μ_2 = mean number of timeouts for students in Option II classrooms.

To determine if the mean number of timeout incidents for students in Option III classrooms is greater than the mean number of timeout incidents for students in Option II classrooms, we test:

$$H_0: \mu_1 - \mu_2 = 0$$
$$H_a: \mu_1 - \mu_2 > 0$$

$$\text{The test statistic is } z = \frac{(\bar{x}_1 - \bar{x}_2) - 0}{\sqrt{\frac{s_1^2}{n_1} + \frac{s_2^2}{n_2}}} = \frac{(102.87 - 78.67) - 0}{\sqrt{\frac{69.33^2}{55} + \frac{59.08^2}{100}}} = 2.19$$

Since α is not given, we will use $\alpha = .05$. The rejection region requires $\alpha = .05$ in the upper tail of the z-distribution. From Table III, Appendix A, $z_{.05} = 16.45$. The rejection region is $z > 1.645$.

Since the observed value of the test statistic falls in the rejection region ($z = 2.19 > 1.645$), H_0 is rejected. There is sufficient evidence to indicate the mean number of timeout incidents for students in Option III classrooms is greater than the mean number of timeout incidents for students in Option II classrooms at $\alpha = .05$.

7.19 Let μ_1 = mean impulsive-sensation seeking score for cocaine abusers and μ_2 = mean impulsive-sensation seeking score for college students.

To determine if there is a difference in the mean impulsive-sensation scores for the two groups, we test:

$$H_0: \mu_1 - \mu_2 = 0$$
$$H_a: \mu_1 - \mu_2 \neq 0$$

$$\text{The test statistic is } z = \frac{(\bar{x}_1 - \bar{x}_2) - 0}{\sqrt{\frac{s_1^2}{n_1} + \frac{s_2^2}{n_2}}} = \frac{(9.4 - 9.5) - 0}{\sqrt{\frac{4.4^2}{450} + \frac{4.4^2}{589}}} = -.36$$

Since no α is given in the problem, we will use $\alpha = .05$. The rejection region requires $\alpha/2 = .05/2 = .025$ in each tail of the z-distribution. From Table III, Appendix A, $z_{.025} = 1.96$. The rejection region is $z < -1.96$ or $z > 1.96$.

Since the observed value of the test statistic does not fall in the rejection region ($z = -.36 \not< -1.96$), H_0 is not rejected. There is insufficient evidence to indicate a difference in the mean impulsive-sensation scores for the two groups at $\alpha = .05$.

Let μ_1 = mean sociability score for cocaine abusers and μ_2 = mean sociability score for college students.

To determine if there is a difference in the mean sociability scores for the two groups, we test:

H_0: $\mu_1 - \mu_2 = 0$
H_a: $\mu_1 - \mu_2 \neq 0$

The test statistic is $z = \dfrac{(\bar{x}_1 - \bar{x}_2) - 0}{\sqrt{\dfrac{s_1^2}{n_1} + \dfrac{s_2^2}{n_2}}} = \dfrac{(10.4 - 12.5) - 0}{\sqrt{\dfrac{4.3^2}{450} + \dfrac{4.0^2}{589}}} = -8.04$

The rejection region is $z < -1.96$ or $z > 1.96$ (from above).

Since the observed value of the test statistic falls in the rejection region ($z = -8.04 < -1.96$), H_0 is rejected. There is sufficient evidence to indicate a difference in the mean sociability scores for the two groups at $\alpha = .05$.

Let μ_1 = mean neuroticism-anxiety score for cocaine abusers and μ_2 = mean neuroticism-anxiety score for college students.

To determine if there is a difference in the mean neuroticism-anxiety scores for the two groups, we test:

H_0: $\mu_1 - \mu_2 = 0$
H_a: $\mu_1 - \mu_2 \neq 0$

The test statistic is $z = \dfrac{(\bar{x}_1 - \bar{x}_2) - 0}{\sqrt{\dfrac{s_1^2}{n_1} + \dfrac{s_2^2}{n_2}}} = \dfrac{(8.6 - 9.1) - 0}{\sqrt{\dfrac{5.1^2}{450} + \dfrac{4.6^2}{589}}} = -1.63$

The rejection region is $z < -1.96$ or $z > 1.96$ (from above).

Since the observed value of the test statistic does not fall in the rejection region ($z = -1.63 \nless -1.96$), H_0 is not rejected. There is insufficient evidence to indicate a difference in the mean neuroticism-anxiety scores for the two groups at $\alpha = .05$.

Let μ_1 = mean aggression-hostility score for cocaine abusers and μ_2 = mean aggression-hostility score for college students.

To determine if there is a difference in the mean aggression-hostility scores for the two groups, we test:

H_0: $\mu_1 - \mu_2 = 0$
H_a: $\mu_1 - \mu_2 \neq 0$

The test statistic is $z = \dfrac{(\bar{x}_1 - \bar{x}_2) - 0}{\sqrt{\dfrac{s_1^2}{n_1} + \dfrac{s_2^2}{n_2}}} = \dfrac{(8.6 - 7.3) - 0}{\sqrt{\dfrac{3.9^2}{450} + \dfrac{4.1^2}{589}}} = 5.21$

The rejection region is $z < -1.96$ or $z > 1.96$ (from above).

Since the observed value of the test statistic falls in the rejection region ($z = 5.21 > 1.96$), H_0 is rejected. There is sufficient evidence to indicate a difference in the mean aggression-hostility scores for the two groups at $\alpha = .05$.

Let μ_1 = mean activity score for cocaine abusers and μ_2 = mean activity score for college students.

To determine if there is a difference in the mean activity scores for the two groups, we test:

$$H_0\colon \mu_1 - \mu_2 = 0$$
$$H_a\colon \mu_1 - \mu_2 \neq 0$$

The test statistic is $z = \dfrac{(\bar{x}_1 - \bar{x}_2) - 0}{\sqrt{\dfrac{s_1^2}{n_1} + \dfrac{s_2^2}{n_2}}} = \dfrac{(11.1 - 8.0) - 0}{\sqrt{\dfrac{3.4^2}{450} + \dfrac{4.1^2}{589}}} = 13.31$

The rejection region is $z < -1.96$ or $z > 1.96$ (from above).

Since the observed value of the test statistic falls in the rejection region ($z = 13.31 > 1.96$), H_0 is rejected. There is sufficient evidence to indicate a difference in the mean activity scores for the two groups at $\alpha = .05$.

7.21 a. Since we want to compare the means of two independent groups, we would use a two sample test of hypothesis for independent samples. Since the sample sizes are small (21 for each), we would use a small sample test.

b. We must assume that both populations being sampled from are approximately normally distributed and that the population variances of the two groups are equal.

c. No. We need to know the common variance of the two groups or an estimate of it.

d. Since the p-value was so large ($p = .79$), H_0 would not be rejected for any reasonable value of α. There is insufficient evidence to indicate a difference in the mean scores for the two groups on the question dealing with clean/sterile gloves at $\alpha \leq .10$.

e. Since the p-value was so small ($p = .02$), H_0 is rejected for any $\alpha > .02$. There is sufficient evidence to indicate a difference in the mean scores for the two groups on the question dealing with a stethoscope at $\alpha > .02$.

7.23 a. Let μ_1 = mean number of cigarettes per week for the treatment group and μ_2 = mean number of cigarettes per week for the control group.

For confidence coefficient .95, $\alpha = .05$ and $\alpha/2 = .025$. From Table IV, Appendix A, with df $= n_1 + n_2 - 2 = 35 + 17 - 2 = 50$, $t_{.025} \approx 2.021$. The confidence interval is:

$$(\bar{x}_1 - \bar{x}_2) \pm t_{.025}\sqrt{s_p^2\left(\frac{1}{n_1} + \frac{1}{n_2}\right)}$$

For Beginning time period:

$$s_p^2 = \frac{(n_1 - 1)s_1^2 + (n_2 - 1)s_2^2}{n_1 + n_2 - 2} = \frac{(35 - 1)71.20^2 + (17 - 1)67.45^2}{35 + 17 - 2} = 4903.06$$

$$\Rightarrow (165.09 - 159.00) \pm 2.021\sqrt{4903.06\left[\frac{1}{35} + \frac{1}{17}\right]}$$

$$\Rightarrow 6.09 \pm 41.835 \Rightarrow (-35.745, 47.925)$$

We are 95% confident that the difference in the mean number of cigarettes smoked per week for the two groups is between -35.745 and 47.925.

For First follow-up period:

$$s_p^2 = \frac{(n_1 - 1)s_1^2 + (n_2 - 1)s_2^2}{n_1 + n_2 - 2} = \frac{(35 - 1)69.08^2 + (17 - 1)66.80^2}{35 + 17 - 2} = 4672.91$$

$$\Rightarrow (105.00 - 157.24) \pm 2.021\sqrt{4672.91\left[\frac{1}{35} + \frac{1}{17}\right]}$$

$$\Rightarrow -52.24 \pm 40.842 \Rightarrow (-93.082, -11.398)$$

We are 95% confident that the difference in the mean number of cigarettes smoked per week for the two groups is between -93.082 and -11.398.

For Second follow-up period:

$$s_p^2 = \frac{(n_1 - 1)s_1^2 + (n_2 - 1)s_2^2}{n_1 + n_2 - 2} = \frac{(35 - 1)69.08^2 + (17 - 1)65.73^2}{35 + 17 - 2} = 4627.53$$

$$\Rightarrow (111.11 - 159.52) \pm 2.021\sqrt{4627.53\left[\frac{1}{35} + \frac{1}{17}\right]}$$

$$\Rightarrow -48.41 \pm 40.643 \Rightarrow (-89.053, -7.767)$$

We are 95% confident that the difference in the mean number of cigarettes smoked per week for the two groups is between -89.053 and -7.767.

For Third follow-up period:

$$s_p^2 = \frac{(n_1 - 1)s_1^2 + (n_2 - 1)s_2^2}{n_1 + n_2 - 2} = \frac{(35 - 1)67.59^2 + (17 - 1)64.41^2}{35 + 17 - 2} = 4434.08$$

$$\Rightarrow (120.20 - 157.88) \pm 2.021\sqrt{4434.08\left[\frac{1}{35} + \frac{1}{17}\right]}$$

$$\Rightarrow -37.68 \pm 39.784 \Rightarrow (-77.464, 2.104)$$

We are 95% confident that the difference in the mean number of cigarettes smoked per week for the two groups is between -77.464 and 2.104.

For Fourth follow-up period:

$$s_p^2 = \frac{(n_1 - 1)s_1^2 + (n_2 - 1)s_2^2}{n_1 + n_2 - 2} = \frac{(35 - 1)74.09^2 + (17 - 1)67.01^2}{35 + 17 - 2} = 5169.65$$

$$\Rightarrow (123.63 - 162.17) \pm 2.021\sqrt{5169.65\left(\frac{1}{35} + \frac{1}{17}\right)}$$

$$\Rightarrow -38.54 \pm 42.958 \Rightarrow (-81.498, 4.418)$$

We are 95% confident that the difference in the mean number of cigarettes smoked per week for the two groups is between −81.498 and 4.418.

b. For each time period, we must make the following assumptions:

1. Both populations being sampled from are normal
2. The two population variances are equal.
3. Independent random samples are selected from each population.

7.25 a. The bacteria counts are probably normally distributed because each count is an average of five measurements from the same specimen.

b. Let μ_1 = mean of the bacteria count for the discharge and μ_2 = mean of the bacteria count upstream. Since we want to test if the mean of the bacteria count for the discharge exceeds the mean of the count upstream, we test:

$H_0: \mu_1 - \mu_2 = 0$
$H_a: \mu_1 - \mu_2 > 0$

c. The test statistic is $t = 1.53$. The two-tailed p-value = .156. Since this is a one-tailed test, $p = .156/2 = .078$.

There is no evidence to indicate that the mean of the bacteria count for the discharge exceeds the mean of the count upstream at $\alpha = .05$.

There is evidence to indicate that the mean of the bacteria count for the discharge exceeds the mean of the count upstream at $\alpha = .10$.

d. We must assume:

1. The mean counts per specimen for each location is normally distributed.
2. The variances of the 2 distributions are equal.
3. Independent and random samples were selected from each population.

7.27 a. Let μ_1 = mean score on the moderate questions for Group A (had 15 difficult questions first) and μ_2 = mean score on the moderate questions for Group B (had 15 easy questions first).

To determine if the students with the easy questions first (Group B) scored better than the students with the difficult questions (Group A), we test:

$H_0: \mu_1 - \mu_2 = 0$
$H_a: \mu_1 - \mu_2 < 0$

b. The p-value = .3248. Since this p-value is not small, H_0 would not be rejected for any reasonable α. There is insufficient evidence to indicate the students with the easy questions first (Group B) scored better than the students with the difficult questions (Group A) at $\alpha \leq .10$.

c. We must assume that the samples were independent.

7.29 $\mu_D = \mu_1 - \mu_2$ = difference in the population means. In testing H_0: $\mu_D = 2$ vs H_a: $\mu_D > 2$, we are testing to see if the difference in the population means ($\mu_1 - \mu_2$) is greater than 2 or not. Or, we are testing to see if the mean for population 1 is more than 2 units larger than the mean for population 2.

7.31 a.

Pair	Difference
1	3
2	2
3	2
4	4
5	0
6	1

$$\bar{x}_D = \frac{\sum x_D}{n} = \frac{12}{6} = 2$$

$$s_D^2 = \frac{\sum x_D^2 - \frac{\left(\sum x_D\right)^2}{n_D}}{n_D - 1} = \frac{\left[34 - \frac{(12)^2}{6}\right]}{5} = 2$$

b. $\mu_D = \mu_1 - \mu_2$

c. For confidence coefficient .95, $\alpha = .05$ and $\alpha/2 = .025$. From Table IV, Appendix A, with df $= n_D - 1 = 6 - 1 = 5$, $t_{.025} = 2.571$. The confidence interval is:

$$\bar{x}_D \pm t_{\alpha/2}\frac{s_D}{\sqrt{n_D}} \Rightarrow 2 \pm 2.571\frac{\sqrt{2}}{\sqrt{6}} \Rightarrow 2 \pm 1.484 \Rightarrow (.516, 3.484)$$

d. H_0: $\mu_D = 0$
H_a: $\mu_D \neq 0$

The test statistic is $t = \dfrac{\bar{x}_D}{s_D/\sqrt{n_D}} = \dfrac{2}{\sqrt{2}/\sqrt{6}} = 3.46$

The rejection region requires $\alpha/2 = .05/2 = .025$ in each tail of the t-distribution with df = $n_D - 1 = 6 - 1 = 5$. From Table IV, Appendix A, $t_{.025} = 2.571$. The rejection region is $t < -2.571$ or $t > 2.571$.

Since the observed value of the test statistic falls in the rejection region ($3.46 > 2.571$), H_0 is rejected. There is sufficient evidence to indicate that the mean difference is different from 0 at $\alpha = .05$.

7.33 a. H_0: $\mu_1 - \mu_2 = 0$
H_a: $\mu_1 - \mu_2 < 0$

The rejection region requires $\alpha = .10$ in the lower tail of the t-distribution with df $= n_D - 1 = 16 - 1 = 15$. From Table IV, Appendix A, $t_{.10} = 1.341$. The rejection region is $t < -1.341$.

b. H_0: $\mu_1 - \mu_2 = 0$
H_a: $\mu_1 - \mu_2 < 0$

The test statistic is $t = \dfrac{\bar{x}_D - 0}{\dfrac{s_D}{\sqrt{n_D}}} = \dfrac{-7 - 0}{\dfrac{\sqrt{64}}{\sqrt{16}}} = -3.5$

The rejection region is $t < -1.341$. (Refer to part a).

Since the observed value of the test statistic falls in the rejection region ($t = -3.5 < -1.341$), H_0 is rejected. There is sufficient evidence to indicate $\mu_1 - \mu_2 < 0$ at $\alpha = .10$.

c. The necessary assumptions are:

1. The population of differences is normal.
2. The differences are randomly selected.

d. For confidence coefficient .90, $\alpha = 1 - .90 = .10$ and $\alpha/2 = .10/2 = .05$. From Table IV, Appendix A, with df $= 15$, $t_{.05} = 1.753$. The confidence interval is:

$$\bar{x}_D \pm t_{.05}\frac{s_D}{\sqrt{n_D}} \Rightarrow -7 \pm 1.753\frac{\sqrt{64}}{\sqrt{16}} \Rightarrow -7 \pm 3.506 \Rightarrow (-10.506, -3.494)$$

We are 90% confident the mean difference is between -10.506 and -3.494.

e. The confidence interval provides more information since it gives an interval of possible values for the difference between the population means.

7.35 a. The variable measured for this experiment is the time needed to match the picture with the sentence.

b. The experimental units are the climbers.

c. Since the same climbers were timed at the base camp and at a camp 5 miles above sea level, the data should be analyzed as a paired experiment.

7.37 Some preliminary calculations:

Number of Homophone Confusions				Number of Homophone Confusions			
Patient	Time 1	Time 2	D = T1 − T2	Patient	Time 1	Time 2	D = T1 − T2
1	5	5	0	11	7	10	−3
2	1	3	−2	12	0	3	−3
3	0	0	0	13	3	9	−6
4	1	1	0	14	5	8	−3
5	0	1	−1	15	7	12	−5
6	2	1	1	16	10	16	−6
7	5	6	−1	17	5	5	0
8	1	2	−1	18	6	3	3
9	0	9	−9	19	9	6	3
10	5	8	−3	20	11	8	3

From the printout, $\bar{x}_D = -1.65$ and $s_D = 3.1999178$.

To determine if Alzheimer's patients show a significant increase in mean homophone confusion errors over time, we test:

H_0: $\mu_D = 0$
H_a: $\mu_D < 0$

The test statistic is $t = \dfrac{\bar{x}_D - 0}{s_D/\sqrt{n_D}} = \dfrac{-1.65}{3.1999/\sqrt{20}} = -2.31$

Since no α was given, we will use $\alpha = .05$. The rejection region requires $\alpha = .05$ in the lower tail of the t-distribution with df $= n_D - 1 = 20 - 1 = 19$. From Table IV, Appendix A, $t_{.05} = 1.729$. The rejection region is $t < -1.729$.

Since the observed value of the test statistic falls in the rejection region ($t = -2.31 < -1.729$), H_0 is rejected. There is sufficient evidence to indicate Alzheimer's patients show a significant increase in mean homophone confusion errors over time at $\alpha = .05$.

7.39 Some preliminary calculations:

Panic Attacks per Week				Panic Attacks per Week			
Patient	Placebo	Inositol	D = Pl − In	Patient	Placebo	Inositol	D = Pl − In
1	0	0	0	12	0	2	−2
2	2	1	1	13	3	1	2
3	0	3	−3	14	3	1	2
4	1	2	−1	15	3	4	−1
5	0	0	0	16	4	2	2
6	10	5	5	17	6	4	2
7	2	0	2	18	15	21	−6
8	6	4	2	19	28	8	20
9	1	1	0	20	30	0	30
10	1	0	1	21	13	0	13
11	1	3	−2				

$$\bar{x}_D = \frac{\sum x_D}{n_D} = \frac{67}{21} = 3.19$$

$$s_D^2 = \frac{\sum x_D^2 - \frac{\left(\sum x_D\right)^2}{n_D}}{n_D - 1} = \frac{\left[1{,}575 - \frac{67^2}{21}\right]}{21 - 1} = 68.0619$$

To determine if there is a difference in the mean number of panic attacks between the placebo and the drug Inositol, we test:

H_0: $\mu_D = 0$
H_a: $\mu_D \neq 0$

The test statistic is $t = \dfrac{\bar{x}_D - 0}{s_D/\sqrt{n_D}} = \dfrac{3.19 - 0}{\sqrt{68.0619}/\sqrt{21}} = 1.77$

Since no α was given, we will use $\alpha = .05$. The rejection region requires $\alpha/2 = .05/2 = .025$ in each tail of the t-distribution with df $= n_D - 1 = 21 - 1 = 20$. From Table IV, Appendix A, $t_{.025} = 2.086$. The rejection region is $t < -2.086$ or $t > 2.086$.

Since the observed value of the test statistic does not fall in the rejection region ($t = 1.77 \not> 2.086$), H_0 is not rejected. There is insufficient evidence to indicate that there is a difference in the mean number of panic attacks between the placebo and the drug Inositol at $\alpha = .05$. There is no evidence that the drug is effective.

7.41 a. Let μ_1 = mean trait for senior year in high school and μ_2 = mean trait during sophomore year of college. Also, let $\mu_D = \mu_1 - \mu_2$. To determine if there is a decrease in the mean self-concept of females between the senior year in high school and the sophomore year in college, we test:

H_0: $\mu_D = 0$
H_a: $\mu_D > 0$

b. Since the same female students were used in each study, the data should be analyzed as paired differences. The samples are not independent.

c. Since the number of pairs is $n_D = 133$, no assumptions about the population of differences is necessary. We still assume that a random sample of differences is selected.

d. Leadership: The p-value $> .05$. There is no evidence to reject H_0 for $\alpha = .05$. There is no evidence to indicate that there is a decrease in mean self-concept of females on leadership between the senior year in high school and the sophomore year in college for $\alpha = .05$.

Popularity: The p-value $< .05$. There is evidence to reject H_0 for $\alpha = .05$. There is evidence to indicate that there is a decrease in mean self-concept of females on popularity between the senior year in high school and the sophomore year in college for $\alpha = .05$.

Intellectual self-confidence: The p-value $< .05$. There is evidence to reject H_0 for $\alpha = .05$. There is evidence to indicate that there is a decrease in mean self-concept of females on intellectual self-concept between the senior year in high school and the sophomore year in college for $\alpha = .05$.

7.43 a.

Pair	Difference
1	−44.48
2	−4.44
3	−48.92
4	−26.69

where μ_D represents the difference in mean pulling strength for the firefighters, (female − male).

$$\bar{x}_D = \frac{\sum x_D}{n_D} = \frac{-124.53}{4} = -31.1325$$

$$s_D^2 = \frac{\sum x_D^2 - \frac{\left(\sum x_D\right)^2}{n_D}}{n_D - 1} = \frac{\left[5103.7065 - \frac{(-124.53)^2}{4}\right]}{3} = 408.9254$$

To determine if there is a difference in mean pulling force between female and male firefighters, we test:

H_0: $\mu_D = 0$
H_a: $\mu_D \neq 0$

The test statistic is $t = \frac{\bar{x}_D - 0}{s_D/\sqrt{n_D}} = \frac{-31.1325 - 0}{\sqrt{408.9254}/\sqrt{4}} = -3.08$

The rejection region requires $\alpha/2 = .05/2 = .025$ in each tail of the t-distribution with df = $n_D - 1 = 4 - 1 = 3$. From Table IV, Appendix A, $t_{.025} = 3.182$. The rejection region is $t < -3.182$ or $t > 3.182$.

Since the observed value of the test statistic does not fall in the rejection region ($-3.08 \nless -3.182$), H_0 is not rejected. There is insufficient evidence to indicate a difference in the mean pulling strength of female and male firefighters at $\alpha = .05$.

b. For confidence coefficient .95, $\alpha = .05$ and $\alpha/2 = .025$. From Table IV, Appendix A, with df = $n_D - 1 = 4 - 1 = 3$, $t_{.025} = 3.182$. The confidence interval is:

$$\bar{x}_D \pm t_{\alpha/2}\frac{s_D}{\sqrt{n_D}} \Rightarrow -31.13 \pm 3.182\frac{\sqrt{408.9254}}{\sqrt{4}} \Rightarrow -31.13 \pm 32.17 \Rightarrow (-63.30, 1.04)$$

We are 95% confident that the difference in the mean pulling strength of firefighters (female − male) is between −63.30 and 1.04.

7.45 a. For confidence coefficient .95, $\alpha = 1 - .95 = .05$ and $\alpha/2 = .05/2 = .025$. From Table III, Appendix A, $z_{.025} = 1.96$.

$$n_1 = n_2 = \frac{(z_{\alpha/2})^2(\sigma_1^2 + \sigma_2^2)}{B^2} = \frac{(1.96)^2(15^2 + 17^2)}{3.2^2} = 192.83 \approx 193$$

b. If the range of each population is 40, we would estimate σ by:

$$\sigma \approx 60/4 = 15$$

For confidence coefficient .99, $\alpha = 1 - .99 = .01$ and $\alpha/2 = .01/2 = .005$. From Table III, Appendix A, $z_{.005} = 2.58$.

$$n_1 = n_2 = \frac{(z_{\alpha/2})^2(\sigma_1^2 + \sigma_2^2)}{B^2} = \frac{(2.58)^2(15^2 + 15^2)}{8^2} = 46.80 \approx 47$$

c. For confidence coefficient .9, $\alpha = 1 - .9 = .1$ and $\alpha/2 = .1/2 = .05$. From Table III, Appendix A, $z_{.05} = 1.645$. For a width of 1, the bound is .5.

$$n_1 = n_2 = \frac{(z_{\alpha/2})^2(\sigma_1^2 + \sigma_2^2)}{B^2} = \frac{(1.645)^2(5.8 + 7.5)}{.5^2} = 143.96 \approx 144$$

7.47 For confidence coefficient .95, $\alpha = .05$ and $\alpha/2 = .025$. From Table III, Appendix A, $z_{.025} = 1.96$.

$$n_1 = n_2 = \frac{(z_{\alpha/2})^2(\sigma_1^2 + \sigma_2^2)}{B^2} = \frac{1.96^2(15^2 + 15^2)}{1^2} = 1728.72 \approx 1729$$

7.49 For confidence coefficient .9, $\alpha = 1 - .9 = .1$ and $\alpha/2 = .1/2 = .05$. From Table III, Appendix A, $z_{.05} = 1.645$.

$$n_1 = n_2 = \frac{(z_{\alpha/2})^2(\sigma_1^2 + \sigma_2^2)}{B^2} = \frac{(1.645)^2(5^2 + 5^2)}{1^2} = 135.3 \Rightarrow 136$$

7.51 From Exercise 7.43, $s_D^2 = 408.9254$. For confidence coefficient .90, $\alpha = .10$ and $\alpha/2 = .10/2 = .05$. From Table III, $z_{.05} = 1.645$.

$$n_D = \frac{(z_{\alpha/2})^2\sigma_D^2}{B^2} = \frac{1.645^2(408.9254)}{15^2} = 4.91 \approx 5$$

Thus, we would need five matched pairs to be able to estimate a difference to within 15 newtons.

7.53 a. The hypotheses are:

H_0: Two sampled populations have identical distributions
H_a: The probability distribution for population A is shifted to the left of that for B

b. First, we rank all the data:

A		B	
Observation	Rank	Observation	Rank
37	8	65	13
40	9	35	6.5
33	3.5	47	11
29	2	52	12
42	10		
33	3.5		
35	6.5		
28	1		
34	5		
	$T_A = 48.5$		$T_B = 42.5$

The test statistic is $T_B = 42.5$ because $n_B < n_A$.

The rejection region is $T_B \geq 39$ from Table V, Appendix A, with $n_A = 9$, $n_B = 4$ and $\alpha = .05$.

Since the observed value of the test statistic falls in the rejection region ($T_B = 42.5 \geq 39$), H_0 is rejected. There is sufficient evidence to indicate the distribution for population B is shifted to the right of the distribution for population A at $\alpha = .05$.

7.55 a. We first rank all the data:

Population 1		Population 2	
Observation	Rank	Observation	Rank
9.0	2	10.1	4
21.1	19	12.0	7
24.8	21	9.2	3
17.2	15	15.8	13
15.6	12	11.1	6
26.9	26	18.2	16
16.5	14	7.0	1
30.1	27	13.6	10
25.6	23	13.5	9
24.6	20	10.3	5
26.0	24	14.2	11
18.7	17	13.2	8
31.1	28		
20.0	18		
25.1	22		
26.1	25		
	$T_A = 313$		$T_B = 93$

To determine whether the probability distribution for population 2 is shifted to the left of that of population 1, we test:

H_0: The two sampled populations have identical probability distributions
H_a: The probability distribution for population 2 is shifted to the left of that of population 1

The test statistic is:

$$z = \frac{T_B - \frac{n_2(n_1 + n_2 + 1)}{2}}{\sqrt{\frac{n_1 n_2(n_1 + n_2 + 1)}{12}}} = \frac{93 - \frac{12(16 + 12 + 1)}{2}}{\sqrt{\frac{16(12)(16 + 12 + 1)}{12}}} = \frac{-81}{21.5407} = -3.76$$

The rejection region requires $\alpha = .05$ in the lower tail of the z-distribution. From Table III, Appendix A, $z_{.05} = 1.645$. The rejection region is $z < -1.645$.

Since the observed value of the test statistic falls in the rejection region ($z = -3.76 < -1.645$), H_0 is rejected. There is sufficient evidence to indicate the probability distribution for population 2 is shifted to the left of that of population 1 at $\alpha = .05$.

b. The p-value is $P(z \leq -3.76) \approx .5 - .5 = 0$ from Table III, Appendix A.

7.57

Neighborhood A	Rank	Neighborhood B	Rank
.850	11	.911	16
1.060	18	.770	3
.910	15	.815	8
.813	7	.748	2
.737	1	.835	9
.880	13	.800	6
.895	14	.793	4
.844	10	.796	5
.965	17		$T_B = 53$
.875	12		
	$T_A = 118$		

a. H_0: The two sampled neighborhoods have identical probability distributions
H_a: The probability distribution for neighborhood A is shifted to the right or left of neighborhood B

The test statistic is $T_B = 53$.

The null hypothesis will be rejected if $T_B \leq T_L$ or $T_B \geq T_U$ where $\alpha = .05$ (two-tailed), $n_1 = 8$ and $n_2 = 10$. From Table V, Appendix A, $T_L = 54$ and $T_U = 98$.

Reject H_0 if $T_B \leq 54$ or $T_B \geq 98$.

129Since $T_B = 53 \leq 54$, we reject H_0 and conclude there is sufficient evidence to indicate neighborhood A is shifted to the right or left of neighborhood B at $\alpha = .05$.

b. The two independent sample t-test is based on the assumptions of:

1. random, independent samples from
2. normally distributed populations with
3. equal variances.

The assumption of normal populations with equal variances would be necessary to use the t-test.

c. 1. The two samples are random and independent.
2. The two probability distributions from which the samples are drawn are continuous.

7.59

Test A	Rank	Test B	Rank
90	12	66	4
71	6	78	8
83	11	50	1
82	10	68	5
75	7	80	9
91	13	60	2
65	3		$T_B = 29$
	$T_A = 62$		

H_0: The probability distributions of scores for tests A and B are identical
H_a: There is a shift in the locations of the probability distributions of scores for tests A and B

The test statistic is $T_B = 29$.

The null hypothesis will be rejected if $T_B \le T_L$ or $T_B \ge T_U$ where $\alpha = .05$ (two-tailed), $n_1 = 6$ and $n_2 = 7$. From Table V, Appendix A, $T_L = 28$ and $T_U = 56$.

Reject H_0 if $T_B \le 28$ or $T_B \ge 56$.

Since $T_B = 29 \not\le 28$ and $T_B = 29 \not\ge 56$, do not reject H_0. There is insufficient evidence to indicate a shift in location for tests A and B at $\alpha = .05$.

7.61

Schedule A	Rank	Schedule B	Rank
15	14	5	6.5
10	13	1	1
5	6.5	2	2.5
7	9.5	8	11
4	5	2	2.5
9	12	6	8
7	9.5	3	4
	$T_A = 69.5$		$T_B = 35.5$

H_0: The probability distributions of the number of hoarded pellets for schedules A and B are identical
H_a: The probability distribution of the number of hoarded pellets for A is shifted either to the right or left of B

The test statistic is $T_A = 69.5$ (or $T_B = 35.5$, since $n_1 = n_2 = 7$)

The rejection region is $T \leq T_L$ or $T \geq T_U$, where $T_L = 37$ is the lower value and $T_U = 68$ is the upper value given by Table V in Appendix A for $\alpha = .05$ two-tailed, with $n_1 = 7$, $n_2 = 7$. The rejection region is $T \leq 37$ or $T \geq 68$.

Since the observed value of the test statistic falls in the rejection region ($69.5 > 68$), H_0 is rejected. There is sufficient evidence to conclude that there is a shift in the distribution of the hoarded pellets at $\alpha = .05$.

7.63 H_0: The probability distributions of eye movement rates are identical for deaf and hearing children
H_a: The probability distribution of eye movement rates for deaf children lies to the right of that for hearing children

From Exercise 7.62, $T_A = 150.5$.

The test statistic is $$z = \frac{T_A - \frac{n_1(n_1 + n_2 + 1)}{2}}{\sqrt{\frac{n_1 n_2(n_1 + n_2 + 1)}{12}}} = \frac{150.5 - \frac{10(10 + 10 + 1)}{2}}{\sqrt{\frac{10(10)(10 + 10 + 1)}{12}}} = 3.44$$

The rejection region requires $\alpha = .05$ in the upper tail of the z-distribution. From Table III, Appendix A, $z_{.05} = 1.645$. The rejection region is $z > 1.645$.

Since the observed value of the test statistic falls in the rejection region ($3.44 > 1.645$), H_0 is rejected. There is sufficient evidence to support the psychologist's claim that deaf children have greater visual acuity than hearing children.

The results agree with those found in Exercise 7.62.

7.65 a. The hypotheses are:

H_0: The two sampled populations have identical probability distributions
H_a: The probability distributions for population A is shifted to the right of that for population B

b. Some preliminary calculations are:

Treatment A	Treatment B	Difference A − B	Rank of Absolute Difference
54	45	9	5
60	45	15	10
98	87	11	7
43	31	12	9
82	71	11	7
77	75	2	2.5
74	63	11	7
29	30	−1	1
63	59	4	4
80	82	−2	2.5
			$T_- = 3.5$

The test statistic is $T_- = 3.5$

The rejection region is $T_- \leq 8$, from Table VI, Appendix A, with $n = 10$ and $\alpha = .025$.

Since the observed value of the test statistic falls in the rejection region ($T_- = 3.5 \leq 8$), H_0 is rejected. There is sufficient evidence to indicate the responses for A tend to be larger than those for B at $\alpha = .025$.

7.67 We assume that the probability distribution of differences is continuous so that the absolute differences will have unique ranks. Although tied (absolute) differences can be assigned average ranks, the number of ties should be small relative to the number of observations to assure validity.

7.69

Pair	Sample from Population 1	Sample from Population 2	Difference	Rank of Absolute Difference
1	8	7	1	1
2	10	1	9	8
3	6	4	2	2.5
4	10	10	0	(eliminated)
5	7	4	3	4
6	8	3	5	6
7	4	6	−2	2.5
8	9	2	7	7
9	8	4	4	5
			Negative rank sum T_-	= 2.5

a. H_0: The probability distributions of the two populations are identical
H_a: The probability distribution of population 1 is shifted to the right of the probability distribution of population 2

The test statistic is $T_- = 2.5$

Reject H_0 if $T_- \leq T_0$ where T_0 is based on $\alpha = .05$ and $n = 8$ (one-tailed):

Reject H_0 if $T_- \leq 6$ (from Table VI, Appendix A).

Since the observed value of the test statistic falls in the rejection region ($T_- = 2.5 < 6$,), H_0 is rejected. There is sufficient evidence to conclude that the probability distribution for population 1 is shifted to the right of the probability distribution for population 2 for $\alpha = .05$.

b. H_0: The probability distributions of the two populations are identical
H_a: The probability distribution of population 1 is shifted either to the right or to the left of the probability distribution of population 2

The test statistic is $T = T_- = 2.5$

Reject H_0 if $T \leq T_0$ where T_0 is based on $\alpha = .05$ and $n = 8$ (two-tailed):

Reject H_0 if $T \leq 4$ (from Table VI, Appendix A).

Since the observed value of the test statistic falls in the rejection region ($T_- = 2.5 < 4$), H_0 is rejected. There is sufficient evidence to conclude that the probability distribution for population 1 is shifted either to the right or to the left of the probability distribution for population 2 for $\alpha = .05$.

7.71 a. This is a paired difference problem.

b. To determine if the problem-solving performance of video teleconferencing groups is superior to face-to-face groups, we test:

H_0: The two sampled populations have identical probability distributions
H_a: The probability distribution for population A (face-to-face) is shifted to the left of that for population B (video teleconferencing)

c. Some preliminary calculations:

Group	Face-To-Face	Video Teleconferencing	Difference	Rank of Absolute Difference
1	65	75	−10	7.5
2	82	80	2	1
3	54	60	−6	6
4	69	65	4	2.5
5	40	55	−15	9
6	85	90	−5	4.5
7	98	98	0	(eliminated)
8	35	40	−5	4.5
9	85	89	−4	2.5
10	70	80	−10	7.5
			Negative rank sum T_- =	41.5
			Positive rank sum T_+ =	3.5

The test statistic is $T_+ = 3.5$.

The null hypothesis will be rejected if $T_+ \le T_0$ where T_0 corresponds to $\alpha = .05$ (one-tailed) and $n = 9$. From Table VI, Appendix A, $T_0 = 8$.

Reject H_0 if $T_+ \le 8$.

Since the observed value of the test statistic falls in the rejection region ($T_+ = 3.5 \le 8$), H_0 is rejected. There is sufficient evidence to indicate that problem-solving performance of video teleconferencing groups is superior to that of groups that interact face-to-face at $\alpha = .05$.

d. p-value $= P(T_+ \le 3.5)$ where $n = 9$ and the test is one-tailed. Using Table VI, locate the appropriate column for n, then find the values in that column that include the test statistic (in this case, 6 and 3). Then read the α level corresponding to these values. Thus,

$.01 < p\text{-value} < .025$

7.73 Some preliminary calculations are:

Twin Pair	School A	School B	Difference	Rank of Absolute Difference
1	65	69	−4	2
2	72	72	0	(Eliminated)
3	86	74	12	4
4	50	52	−2	1
5	60	47	13	5
6	81	72	9	3
				$T_+ = 12$
				$T_- = 3$

To determine whether there is a difference in the probability distributions of achievement scores at the two schools, we test:

H_0: The achievement test scores are the same for both junior high schools
H_a: The achievement test scores of junior high school A are shifted to the right or left of those for junior high school B

The test statistic is $T_- = 3$.

Reject H_0 if $T \leq T_0$ where T_0 is based on $n = 5$ and $\alpha = .10$ (two-tailed):

Reject H_0 if $T \leq 1$ (from Table VI, Appendix A).

Since the observed value of the test statistic does not fall in the rejection region, ($T = 3 \nleq 1$), H_0 is not rejected. There is no difference in the probability distributions of achievement test scores at the two schools at $\alpha = .10$.

7.75 Some preliminary calculations are:

Charcoal Canister No.	PCHD	EERF	Difference	Rank of Absolute Difference
71	1,709.79	1,479.0	230.79	11
58	357.17	257.8	99.37	2
84	1,150.94	1,287.0	−136.06	5
91	1,572.69	1,395.0	177.69	9
44	558.33	416.5	141.83	8
43	4,132.28	3,993.0	139.28	7
79	1,489.86	1,351.0	138.86	6
61	3,017.48	1,813.0	1,204.48	15
85	393.55	187.7	205.85	10
46	880.84	630.4	250.44	12
4	2,996.49	3,707.0	−710.51	14
20	2,367.40	2,791.0	−423.60	13
36	599.84	706.8	−106.96	3
42	538.37	618.5	−80.13	1
55	2,770.23	2,639.0	131.23	4
				$T_+ = 84$
				$T_- = 36$

To determine if one of the measuring facilities tends to read higher or lower than the other, we test:

H_0: The exhalation rate measurements for the two facilities are the same
H_a: The exhalation rate measurements by PCHD are shifted to the right or left of those by EERF

The test statistic is $T_- = 36$.

Reject H_0 if $T \leq T_0$ where T_0 is based on $n = 15$ and $\alpha = .05$ (two-tailed):

Reject H_0 if $T \leq 25$ (from Table VI, Appendix A).

Since the observed value of the test statistic does not fall in the rejection region ($T = 36 \nleq 25$), H_0 is not rejected. There is insufficient evidence to indicate a difference in the exhalation rate measurements for the two facilities at $\alpha = .05$.

7.77 a. $P(F \leq 3.48) = 1 - .05 = .95$ using Table VIII, Appendix A, with $\nu_1 = 5$ and $\nu_2 = 9$

b. $P(F > 3.09) = .01$ using Table X, Appendix A, with $\nu_1 = 15$ and $\nu_2 = 20$

c. $P(F > 2.40) = .05$ using Table VIII, Appendix A, with $\nu_1 = 15$ and $\nu_2 = 15$

d. $P(F \leq 1.83) = 1 - .10 = .90$ using Table VII, Appendix A, with $\nu_1 = 8$ and $\nu_2 = 40$

7.79 a. For diagram **a**,

$$\bar{x}_1 = \frac{\sum x_1}{n} = \frac{7 + 8 + 9 + 9 + 10 + 11}{6} = \frac{54}{6} = 9$$

$$\bar{x}_2 = \frac{\sum x_2}{n} = \frac{12 + 13 + 14 + 14 + 15 + 16}{6} = \frac{84}{6} = 14$$

For diagram **b**,

$$\bar{x}_1 = \frac{\sum x_1}{n} = \frac{5 + 5 + 7 + 11 + 13 + 13}{6} = \frac{54}{6} = 9$$

$$\bar{x}_2 = \frac{\sum x_2}{n} = \frac{10 + 10 + 12 + 16 + 18 + 18}{6} = \frac{84}{6} = 14$$

b. For diagram **a**,

$$\text{SST} = \sum_{i=1}^{2} n_i(\bar{x}_i - \bar{x})^2 = 6(9 - 11.5)^2 + 6(14 - 11.5)^2 = 75$$

$$(\bar{x} = \frac{\sum x}{n} = \frac{54 + 84}{12} = 11.5)$$

For diagram **b**,

$$\text{SST} = \sum_{i=1}^{2} n_i(\bar{x}_i - \bar{x})^2 = 6(9 - 11.5)^2 + 6(14 - 11.5)^2 = 75$$

c. For diagram **a**,

$$s_1^2 = \frac{\sum x_1^2 - \frac{\left(\sum x_1\right)^2}{n_1}}{n_1 - 1} = \frac{496 - \frac{54^2}{6}}{6 - 1} = 2$$

$$s_2^2 = \frac{\sum x_2^2 - \frac{\left(\sum x_2\right)^2}{n_2}}{n_2 - 1} = \frac{1186 - \frac{84^2}{6}}{6 - 1} = 2$$

$$\text{SSE} = (n_1 - 1)s_1^2 + (n_2 - 1)s_2^2 = (6 - 1)2 + (6 - 1)2 = 20$$

For diagram **b**,

$$s_1^2 = \frac{\sum x_1^2 - \frac{\left(\sum x_1\right)^2}{n_1}}{n_1 - 1} = \frac{558 - \frac{54^2}{6}}{6 - 1} = 14.4$$

$$s_2^2 = \frac{\sum x_2^2 - \frac{\left(\sum x_2\right)^2}{n_2}}{n_2 - 1} = \frac{1248 - \frac{84^2}{6}}{6 - 1} = 14.4$$

$$\text{SSE} = (n_1 - 1)s_1^2 + (n_2 - 1)s_2^2 = (6 - 1)14.4 + (6 - 1)14.4 = 144$$

d. For diagram **a**, SS(Total) = SST + SSE = 75 + 20 = 95

$$\text{SST is } \frac{\text{SST}}{\text{SS(Total)}} \times 100\% = \frac{75}{95} \times 100\% = 78.95\% \text{ of SS(Total)}$$

For diagram **b**, SS(Total) = SST + SSE = 75 + 144 = 219

$$\text{SST is } \frac{\text{SST}}{\text{SS(Total)}} \times 100\% = \frac{75}{219} \times 100\% = 34.25\% \text{ of SS(Total)}$$

e. For diagram **a**, $\text{MST} = \frac{\text{SST}}{p - 1} = \frac{75}{2 - 1} = 75$

$$\text{MSE} = \frac{\text{SSE}}{n - p} = \frac{20}{12 - 2} = 2 \qquad F = \frac{\text{MST}}{\text{MSE}} = \frac{75}{2} = 37.5$$

For diagram **b**, $\text{MST} = \frac{\text{SST}}{p - 1} = \frac{75}{2 - 1} = 75$

$$\text{MSE} = \frac{\text{SSE}}{n - p} = \frac{144}{12 - 2} = 14.4 \qquad F = \frac{\text{MST}}{\text{MSE}} = \frac{75}{14.4} = 5.21$$

f. The rejection region for both diagrams requires $\alpha = .05$ in the upper tail of the F-distribution with numerator df $= p - 1 = 2 - 1 = 1$ and denominator df $= n - p = 12 - 2 = 10$. From Table VIII, Appendix A, $F_{.05} = 4.96$. The rejection region is $F > 4.96$.

For diagram **a**, the observed value of the test statistic falls in the rejection region ($F = 37.5 > 4.96$). Thus, H_0 is rejected. There is sufficient evidence to indicate the samples were drawn from populations with different means at $\alpha = .05$.

For diagram **b**, the observed value of the test statistic falls in the rejection region ($F = 5.21 > 4.96$). Thus, H_0 is rejected. There is sufficient evidence to indicate the samples were drawn from populations with different means at $\alpha = .05$.

g. We must assume both populations are normally distributed with common variances.

7.81 Refer to Exercise 7.79, the ANOVA table is:

For diagram **a**:

Source	df	SS	MS	F
Treatment	1	75	75	37.5
Error	10	20	2	
Total	11	95		

For diagram **b**:

Source	df	SS	MS	F
Treatment	1	75	75	5.21
Error	10	144	14.4	
Total	11	219		

7.83 a. The number of treatments is $3 + 1 = 4$. The total sample size is $37 + 1 = 38$.

b. To determine if the treatment means differ, we test:

H_0: $\mu_1 = \mu_2 = \mu_3 = \mu_4$
H_a: At least two treatment means differ

The test statistic is $F = 14.80$.

The rejection region requires $\alpha = .01$ in the upper tail of the F-distribution with $\nu_1 = p - 1 = 4 - 1 = 3$ and $\nu_2 = n - p = 38 - 4 = 34$. From Table X, Appendix A, $F_{.01} \approx 4.51$. The rejection region is $F > 4.51$.

Since the observed value of the test statistic falls in the rejection region ($F = 14.80 > 4.51$), H_0 is rejected. There is sufficient evidence to indicate differences among the treatment means at $\alpha = .01$.

c. We need the sample means to compare specific pairs of treatment means.

7.85 a. To determine whether the mean scores of the four groups differ, we test:

H_0: $\mu_1 = \mu_2 = \mu_3 = \mu_4$
H_a: At least two treatment means differ

where μ_i represents the mean of the ith group.

For the variable "Infrequency":

The test statistic is $F = 155.8$.

The rejection region requires $\alpha = .05$ in the upper tail of the F-distribution with $\nu_1 = p - 1 = 4 - 1 = 3$ and $\nu_2 = n - p = 278 - 4 = 274$. Using Table VIII, Appendix A, $F_{.05} \approx 2.60$. The rejection region is $F > 2.60$.

Since the observed value of the test statistic falls in the rejection region ($F = 155.5 > 2.60$), H_0 is rejected. There is sufficient evidence to indicate the mean scores on the "Infrequency" variable differ among the four groups at $\alpha = .05$.

For the variable "Obvious":

The test statistic is $F = 49.7$.

The rejection region is $F > 2.60$. (See above.)

Since the observed value of the test statistic falls in the rejection region ($F = 49.7 > 2.60$), H_0 is rejected. There is sufficient evidence to indicate the mean scores on the "Obvious" variable differ among the four groups at $\alpha = .05$.

For the variable "Subtle":

The test statistic is $F = 10.3$.

The rejection region is $F > 2.60$. (See above.)

Since the observed value of the test statistic falls in the rejection region ($F = 10.3 > 2.60$), H_0 is rejected. There is sufficient evidence to indicate the mean scores on the "Subtle" variable differ among the four groups at $\alpha = .05$.

For the variable "Obvious-Subtle":

The test statistic is $F = 45.4$.

The rejection region is $F > 2.60$. (See above.)

Since the observed value of the test statistic falls in the rejection region ($F = 45.4 > 2.60$), H_0 is rejected. There is sufficient evidence to indicate the mean scores on the "Obvious-Subtle" variable differ among the four groups at $\alpha = .05$.

For the variable "Dissimulation":

The test statistic is $F = 39.1$.

The rejection region is $F > 2.60$. (See above.)

Since the observed value of the test statistic falls in the rejection region ($F = 39.1 > 2.60$), H_0 is rejected. There is sufficient evidence to indicate the mean scores on the "Dissimulation" variable differ among the four groups at $\alpha = .05$.

b. No. No information is provided on the sample means. The test of hypotheses performed in part **a** just indicate differences exist, but do not indicate where. Further analysis would be required.

7.87 a. If the samples represent random and independent selections from the populations associated with the four hair colors, then a completely randomized design has been used.

b. Let μ_1, μ_2, μ_3, and μ_4 represent the mean pain threshold scores for all people with light blond, dark blond, light brunette, and dark brunette hair, respectively.

H_0: $\mu_1 = \mu_2 = \mu_3 = \mu_4$
H_a: At least two means differ

Test statistic: $F = \dfrac{\text{MST}}{\text{MSE}} = 6.79$

The rejection region requires $\alpha = .05$ in the upper tail of the F-distribution with $\nu_1 = p - 1 = 4 - 1 = 3$ and $\nu_2 = n - p = 19 - 4 = 15$. From Table VIII, Appendix A, $F_{.05} = 3.29$. The rejection region is $F > 3.29$.

Since the observed value of the test statistic falls in the rejection region ($F = 6.79 > 3.29$), H_0 is rejected. There is sufficient evidence to indicate that at least one of the mean pain thresholds is different from the others at $\alpha = .05$.

c. The p-value is given on the output under "Pr > F" as 0.0041. Because this value is so small, this is a strong indication that the mean pain thresholds differ.

d. The assumptions necessary to assure the validity of the test are as follows:

1. The probability distributions of the pain threshold scores is normal for each hair color.
2. The variances of the probability distributions of the pain threshold scores for each hair color are equal.
3. The samples are selected randomly and independently.

7.89 a. A completely randomized design was used.

b. To determine if the mean concern ratings differ for at least two education levels, we test:

H_0: $\mu_1 = \mu_2 = \mu_3 = \mu_4 = \mu_5$
H_a: At least two treatment means differ

where μ_i represents the mean concern rating of the ith education level.

The test statistic is $F = 3.298$.

The rejection region requires $\alpha = .05$ in the upper tail of the F-distribution with $\nu_1 = p - 1 = 5 - 1 = 4$ and $\nu_2 = n - p = 315 - 5 = 310$. From Table VIII, Appendix A, $F_{.05} \approx 2.37$. The rejection region is $F > 2.37$.

Since the observed value of the test statistic falls in the rejection region ($F = 3.298 > 2.37$), H_0 is rejected. There is sufficient evidence to indicate a difference in the mean concern ratings among the 5 education levels at $\alpha = .05$.

7.91 a. $s_p^2 = \dfrac{(n_1 - 1)s_1^2 + (n_1 - 1)s_2^2}{n_1 + n_2 - 2} = \dfrac{11(74.2) + 13(60.5)}{12 + 14 - 2} = 66.7792$

H_0: $\mu_1 - \mu_2 = 0$
H_a: $\mu_1 - \mu_2 > 0$

The test statistic is $t = \dfrac{(\bar{x}_1 - \bar{x}_2) - 0}{\sqrt{s_p^2\left(\dfrac{1}{n_1} + \dfrac{1}{n_2}\right)}} = \dfrac{(17.8 - 15.3) - 0}{\sqrt{66.7792\left(\dfrac{1}{12} + \dfrac{1}{14}\right)}} = .78$

The rejection region requires $\alpha = .05$ in the upper tail of the t-distribution with df $= n_1 + n_2 - 2 = 12 + 14 - 2 = 24$. From Table IV, Appendix A, for df = 24, $t_{.05} = 1.711$. The rejection region is $t > 1.711$.

Since the observed value of the test statistic does not fall in the rejection region ($0.78 \not> 1.711$), H_0 is not rejected. There is insufficient evidence to indicate that $\mu_1 > \mu_2$ at $\alpha = .05$.

b. For confidence coefficient .99, $\alpha = .01$ and $\alpha/2 = .01/2 = .005$. From Table IV, Appendix A, with df $= n_1 + n_2 - 2 = 12 + 14 - 2 = 24$, $t_{.005} = 2.797$. The confidence interval is:

$$(\bar{x}_1 - \bar{x}_2) \pm t_{.005}\sqrt{s_p^2\left(\frac{1}{n_1} + \frac{1}{n_2}\right)} \Rightarrow (17.8 - 15.3) \pm 2.797\sqrt{66.7792\left(\frac{1}{12} + \frac{1}{14}\right)}$$

$$\Rightarrow 2.50 \pm 8.99 \Rightarrow (-6.49,\ 11.49)$$

c. For confidence coefficient .99, $\alpha = .01$ and $\alpha/2 = .01/2 = .005$. From Table III, Appendix A, $z_{.005} = 2.58$.

$$n_1 = n_2 = \frac{(z_{\alpha/2})^2(\sigma_1^2 + \sigma_2^2)}{B^2} = \frac{(2.58)^2(74.2 + 60.5)}{2^2} = 224.15 \approx 225$$

7.93 a. This is a paired difference experiment.

Pair	Difference (Pop. 1 − Pop. 2)
1	6
2	4
3	4
4	3
5	2

$$\bar{x}_D = \frac{\sum x_D}{n_D} = \frac{19}{5} = 3.8 \qquad s_D^2 = \frac{\sum x_D^2 - \frac{\left(\sum x_D\right)^2}{n_D}}{n_D - 1} = \frac{81 - \frac{19^2}{5}}{5 - 1} = 2.2$$

$$s_D = \sqrt{2.2} = 1.4832$$

H_0: $\mu_D = 0$
H_a: $\mu_D \neq 0$

The test statistic is $t = \dfrac{\bar{x}_D - 0}{s_D/\sqrt{n_D}} = \dfrac{3.8 - 0}{1.4832/\sqrt{5}} = 5.73$

The rejection region requires $\alpha/2 = .05/2 = .025$ in each tail of the t-distribution with df $= n - 1 = 5 - 1 = 4$. From Table IV, Appendix A, $t_{.025} = 2.776$. The rejection region is $t < -2.776$ or $t > 2.776$.

Since the observed value of the test statistic falls in the rejection region ($5.73 > 2.776$), H_0 is rejected. There is sufficient evidence to indicate that the population means are different at $\alpha = .05$.

b. For confidence coefficient .95, $\alpha = .05$ and $\alpha/2 = .025$. Therefore, we would use the same t value as above, $t_{.025} = 2.776$. The confidence interval is:

$$\bar{x}_D \pm t_{\alpha/2}\frac{s_D}{\sqrt{n_D}} \Rightarrow 3.8 \pm 3.8 \pm 2.776\frac{1.4832}{\sqrt{5}} \Rightarrow 3.8 \pm 1.84 \Rightarrow (1.96, 5.64)$$

c. The sample of differences must be randomly selected from a population of differences which has a normal distribution.

7.95 The appropriate test for two independent samples is the Wilcoxon Rank Sum test. Some preliminary calculations are:

Sample 1	Rank	Sample 2	Rank
1.2	4	1.5	6
1.9	8.5	1.3	5
.7	1	2.9	12
2.5	10	1.9	8.5
1.0	2	2.7	11
1.8	7	3.5	13
1.1	3		
	$R_A = 35.5$		$R_B = 55.5$

To determine if there is a difference between the locations of the probability distributions, we test:

H_0: The two sampled populations have identical probability distributions
H_a: The probability distribution for population A is shifted to the left or right of that for B

The test statistic is $T_B = 55.5$.

Reject H_0 if $T_B \leq T_L$ or $T_B \geq T_U$ where $\alpha = .05$ (two-tailed), $n_1 = 7$ and $n_2 = 6$:

Reject H_0 if $T_B \leq 28$ or $T_B \geq 56$ (from Table V, Appendix A).

Since $T_B = 55.5 \nleq 28$ and $T_B = 55.5 \ngeq 56$, do not reject H_0. There is insufficient evidence to indicate a difference between the locations of the probability distributions for the sampled populations at $\alpha = .05$.

7.97 a. SSE $=$ SSTot $-$ SST $= 62.55 - 36.95 = 25.60$

df Treatment $= p - 1 = 4 - 1 = 3$
df Error $= n - p = 20 - 4 = 16$
df Total $= n - 1 = 20 - 1 = 19$

$$\text{MST} = \text{SST/df} = \frac{36.95}{3} = 12.32$$

$$\text{MSE} = \text{SSE/df} = \frac{25.60}{16} = 1.60$$

$$F = \frac{\text{MST}}{\text{MSE}} = \frac{12.32}{1.60} = 7.70$$

The ANOVA table:

Source	df	SS	MS	F
Treatment	3	36.95	12.32	7.70
Error	16	25.60	1.60	
Total	19	62.55		

b. To determine if there is a difference in the treatment means, we test:

H_0: $\mu_1 = \mu_2 = \mu_3 = \mu_4$
H_a: At least two of the means differ

where the μ_i represents the mean for the ith treatment.

The test statistic is $F = \frac{\text{MST}}{\text{MSE}} = 7.70$

The rejection region requires $\alpha = .10$ in the upper tail of the F-distribution with $\nu_1 = (p - 1) = (4 - 1) = 3$ and $\nu_2 = (n - p) = (20 - 4) = 16$. From Table VII, Appendix A, $F_{.10} = 2.46$. The rejection region is $F > 2.46$.

Since the observed value of the test statistic falls in the rejection region ($F = 7.70 > 2.46$), H_0 is rejected. There is sufficient evidence to conclude that at least two of the means differ at $\alpha = .10$.

c. $\bar{x}_4 = \frac{\sum x_4}{n_4} = \frac{57}{5} = 11.4$

For confidence level .90, $\alpha = .10$ and $\alpha/2 = .10/2 = .05$. From Table IV, Appendix A, with df $= 16$, $t_{.05} = 1.746$. The confidence interval is:

$$\bar{x}_4 \pm t_{.05} \cdot \sqrt{\text{MSE}/n_4} \Rightarrow 11.4 \pm 1.746 \cdot \sqrt{1.6/5} \Rightarrow 11.4 \pm .99 \Rightarrow (10.41, 12.39)$$

7.99 a. This is a paired difference experiment. Each person in the experiment is responding to pattern 1 and pattern 2.

b. For confidence coefficient .90, $\alpha = .10$ and $\alpha/2 = .05$. From Table IV, Appendix A, with df $= n_D - 1 = 15 - 1 = 14$, $t_{.05} = 1.761$. The confidence interval is:

$$\bar{x}_D \pm t_{\alpha/2} \frac{s_D}{\sqrt{n_D}} \Rightarrow .24 \pm 1.761 \frac{.16}{\sqrt{15}} \Rightarrow .24 \pm .073 \Rightarrow (.167, .313)$$

We are 90% confident that the difference in the mean response for pattern 1 is greater than the mean response for pattern 2 by between .167 and .313

c. To determine if there is sufficient evidence to indicate a difference between the mean responses for the two patterns, we test:

H_0: $\mu_D = 0$
H_a: $\mu_D \neq 0$

where μ_D represents the difference in mean response, (pattern 1 − pattern 2).

The test statistic is $t = \dfrac{\bar{x}_D - 0}{s_D/\sqrt{n_D}} = \dfrac{.24 - 0}{.16/\sqrt{15}} = 5.81$

The rejection region requires $\alpha/2 = .10/2 = .05$ in each tail of the t-distribution with df = $n_D - 1 = 15 - 1 = 14$. From Table IV, Appendix A, $t_{.05} = 1.761$. The rejection region is $t < -1.761$ or $t > 1.761$.

Since the observed value of the test statistic falls in the rejection region ($5.81 > 1.761$), H_0 is rejected. There is sufficient evidence to indicate a difference in the mean responses to the patterns at $\alpha = .10$. This agrees with the interpretation of the confidence interval.

d. In order for the above procedures to be valid, we must be able to assume:

1. The differences in the pupillometer measurements have a relative frequency distribution that is approximately normal.
2. The differences are randomly selected.

7.101 a. Let μ_1 = mean rating of concern about product tampering for males and μ_2 = mean rating of concern about product tampering for females. To determine whether a difference exists in the mean level of concern about product tampering between males and females, we test:

H_0: $\mu_1 - \mu_2 = 0$
H_a: $\mu_1 - \mu_2 \neq 0$

b. The test statistic is $z = -2.69$ and the p-value $= .0072$. Since the p-value is so small, there is evidence to reject H_0. There is sufficient evidence to indicate a difference exists in the mean level of concern about product tampering between males and females for $\alpha > .0072$.

c. We must assume the sample sizes were sufficiently large so that the Central Limit Theorem applies. We must also assume that we selected two random and independent samples from the two populations.

7.103 Let μ_1 = mean rating for music majors and μ_2 = mean rating of nonmusic majors.

For confidence coefficient .95, $\alpha = .05$ and $\alpha/2 = .05/2 = .025$. From Table III, Appendix A, $z_{.025} = 1.96$. The confidence interval is:

$$(\bar{x}_1 - \bar{x}_2) \pm z_{.025}\sqrt{\frac{s_1^2}{n_1} + \frac{s_2^2}{n_2}} \Rightarrow (4.26 - 4.59) \pm 1.96\sqrt{\frac{.81^2}{100} + \frac{.78^2}{100}}$$
$$\Rightarrow -.33 \pm .220 \Rightarrow (-.550, -.110)$$

We are 95% confident that the difference in the mean rating between music majors and nonmusic majors is between −.55 and −.11.

7.105 There is probably much variation in the abilities of mentally impaired subjects. To control some of this variation, a paired experiment is used. We would assume that within each pair (matched by IQ), there is little variation. We would assume that the variation between the pairs is great. By identifying and controlling for one source of variation, we will be more precise in our estimation of the difference in the mean scores between the groups.

7.107 a. df Total $= n - 1 = 167 - 1 = 166$; df Treatment $= p - 1 = 4 - 1 = 3$

SSE = SStot − SST = 120.62 − 21.31 = 99.31

MSE = SSE/df = 99.31/163 = .6093

F = MST/MSE = 7.10/.6093 = 11.65

Source	df	SS	MS	F
Staff Affil	3	21.31	7.10	11.65
Error	163	99.31	0.6093	
Total	166	120.62		

b. To determine if there is a difference among the mean levels of conflict perceived by the four groups of jail staff members, we test:

H_0: $\mu_1 = \mu_2 = \mu_3 = \mu_4$
H_a: At least one treatment mean differs

where μ_i is the mean level of perceived conflict for group i.

The test statistic is $F = 11.65$.

The rejection region requires $\alpha = .05$ in the upper tail of the F-distribution with $\nu_1 = p - 1 = 4 - 1 = 3$ and $\nu_2 = n - p = 167 - 4 = 163$. From Table VIII, Appendix A, $F_{.05} \approx 2.65$. The rejection region is $F > 2.65$.

Since the observed value of the test statistic falls in the rejection region ($F = 11.65 > 2.65$), H_0 is rejected. There is sufficient evidence to indicate a difference among the mean levels of conflict perceived by the four groups of jail staff members at $\alpha = .05$.

7.109 If the p-value is less than α, reject H_0. Otherwise, do not reject H_0.

a. p-value $= .0429 < .05 \Rightarrow$ Reject H_0

b. p-value $= .1984 \nless .05 \Rightarrow$ Do not reject H_0

c. p-value $= .0001 < .05 \Rightarrow$ Reject H_0

d. p-value $= .0344 < .05 \Rightarrow$ Reject H_0

e. p-value $= .0545 \nless .05 \Rightarrow$ Do not reject H_0

f. p-value $= .9633 \nless .05 \Rightarrow$ Do not reject H_0

g. It would be necessary to be able to assume that the two samples were independently and randomly taken from normal populations with the same variance.

7.111 a. For each of the three measures, let μ_1 = mean score for males seeing the first advertisement and μ_2 = mean score for males seeing the second advertisement. Also, let $\mu_D = \mu_1 - \mu_2$. To determine whether the first ad will be more effective when shown to males, we test:

$$H_0: \mu_D = 0$$
$$H_a: \mu_D > 0$$

b. This experiment was a paired difference experiment. Each male was shown both advertisements.

c. Attitude towards the Advertisement:

The p-value = .091. There is no evidence to reject H_0 for $\alpha = .05$. There is no evidence to indicate the first ad will be more effective when shown to males for $\alpha = .05$. There is evidence to reject H_0 for $\alpha = .10$. There is evidence to indicate the first ad will be more effective when shown to males for $\alpha = .10$.

Attitude toward Brand of Soft Drink:

The p-value = .032. There is evidence to reject H_0 for $\alpha > .032$. There is evidence to indicate the first ad will be more effective when shown to males for $\alpha > .032$.

Intention to Purchase the Soft Drink:

The p-value = .050. There is no evidence to reject H_0 for $\alpha = .05$. There is no evidence to indicate the first ad will be more effective when shown to males for $\alpha = .05$. There is evidence to reject H_0 for $\alpha > .050$. There is evidence to indicate the first ad will be more effective when shown to males for $\alpha > .050$.

d. We must assume that the sample of differences is randomly selected.

7.113 a. Let μ_1 = mean intensity of parental punishment for single mothers and μ_2 = mean intensity of parental punishment for married mothers.

$$s_p^2 = \frac{(n_1 - 1)s_1^2 + (n_2 - 1)s_2^2}{n_1 + n_2 - 2} = \frac{7(14.8)^2 + 5(13.69)^2}{8 + 6 - 2} = 205.8634$$

For confidence coefficient .95, $\alpha = .05$ and $\alpha/2 = .025$. From Table IV, Appendix A, with df $= n_1 + n_2 - 2 = 8 + 6 - 2 = 12$, $t_{.025} = 2.179$. The confidence interval is:

$$(\bar{x}_1 - \bar{x}_2) \pm t_{.025}\sqrt{s_p^2\left(\frac{1}{n_1} + \frac{1}{n_2}\right)}$$

$$(70.75 - 77.33) \pm 2.179\sqrt{205.8634\left(\frac{1}{8} + \frac{1}{6}\right)}$$

$$-6.58 \pm 16.88 \Rightarrow (-23.46, 10.30)$$

We are 95% confident the difference between the mean intensity of parental punishment scores for single and married mothers is between −23.46 and 10.30.

b. Since this confidence indicates that 0 is a reasonable value, it does not support an inference that the true means for single and married mothers differ at $\alpha = .05$.

c. To determine if there is sufficient evidence to indicate a difference between the intensity of parental punishment between single and married mothers, we test:

$H_0\colon \mu_1 - \mu_2 = 0$
$H_a\colon \mu_1 - \mu_2 \neq 0$

where μ_1 = mean parental punishment intensity score for single mothers
μ_2 = mean parental punishment intensity score for married mothers

The test statistic is $t = \dfrac{(\bar{x}_1 - \bar{x}_2) - D_0}{\sqrt{s_p^2\left(\dfrac{1}{n_1} + \dfrac{1}{n_2}\right)}} = \dfrac{(70.75 - 77.33)}{\sqrt{205.8634\left(\dfrac{1}{8} + \dfrac{1}{6}\right)}} = -.85$

The rejection region requires $\alpha/2 = .05/2 = .025$ in each tail of the t-distribution with df = $n_1 + n_2 - 2 = 8 + 6 - 2 = 12$. From Table IV, Appendix A, $t_{.025} = 2.179$. The rejection region is $t < -2.179$ or $t > 2.179$.

Since the observed value of the test statistic does not fall in the rejection region ($-.85 \not< 2.179$), H_0 is not rejected. There is insufficient evidence to indicate that single and married mothers differ in their punishment practices, at $\alpha = .05$.

d. In order for the above procedures to be valid, we must be able to assume:

1. The intensity scores of parental punishment for single mothers and married mothers have relative frequency distributions that are approximately normal.
2. The variances for these two populations are equal.
3. The samples are randomly and independently selected.

Comparing Population Proportions

Chapter 8

8.1 A binomial experiment has the following five characteristics:

1. The experiment consists of n identical trials.
2. There are only two possible outcomes on each trial (Success/Failure).
3. The probability of S on any trial, p, is a constant.
4. The trials are independent.
5. The random variable is the number of S's in n trials.

8.3 From Section 6.5, it was given that the distribution of $\hat{p}$ is approximately normal if the interval $\hat{p} \pm 3\sigma_{\hat{p}}$ does not contain 0 or 1.

a. $\sigma_{\hat{p}_1} = \sqrt{\dfrac{\hat{p}_1\hat{q}_1}{n_1}} = \sqrt{\dfrac{.5(.5)}{10}} = .158 \quad \hat{p}_1 \pm 3\sigma_{\hat{p}_1} \Rightarrow .5 \pm 3(.158) \Rightarrow .5 \pm .474 \Rightarrow (.026, .974)$

$\sigma_{\hat{p}_2} = \sqrt{\dfrac{\hat{p}_2\hat{q}_2}{n_2}} = \sqrt{\dfrac{.5(.5)}{12}} = .144 \quad \hat{p}_2 \pm 3\sigma_{\hat{p}_2} \Rightarrow .5 \pm 3(.144) \Rightarrow .5 \pm .432 \Rightarrow (.068, .932)$

Yes. Neither interval contains 0.

b. $\sigma_{\hat{p}_1} = \sqrt{\dfrac{.1(.9)}{10}} = .095 \quad \hat{p}_1 \pm 3\sigma_{\hat{p}_1} \Rightarrow .1 \pm 3(.095) \Rightarrow .1 \pm .285 \Rightarrow (-.185, .385)$

$\sigma_{\hat{p}_2} = \sqrt{\dfrac{.08(.92)}{12}} = .078 \quad \hat{p}_2 \pm 3\sigma_{\hat{p}_2} \Rightarrow .08 \pm 3(.078) \Rightarrow .08 \pm .234 \Rightarrow (-.154, .314)$

No. Both intervals contains 0.

c. $\sigma_{\hat{p}_1} = \sqrt{\dfrac{.2(.8)}{30}} = .073 \quad \hat{p}_1 \pm 3\sigma_{\hat{p}_1} \Rightarrow .2 \pm 3(.073) \Rightarrow .2 \pm .219 \Rightarrow (-.019, .419)$

$\sigma_{\hat{p}_2} = \sqrt{\dfrac{.3(.7)}{30}} = .084 \quad \hat{p}_2 \pm 3\sigma_{\hat{p}_2} \Rightarrow .3 \pm 3(.084) \Rightarrow .3 \pm .252 \Rightarrow (.048, .552)$

No. The interval $\hat{p}_1 \pm 3\sigma_{\hat{p}_1}$ contains 0.

d. $\sigma_{\hat{p}_1} = \sqrt{\dfrac{.05(.95)}{100}} = .022 \quad \hat{p}_1 \pm 3\sigma_{\hat{p}_1} \Rightarrow .05 \pm 3(.022) \Rightarrow .05 \pm .066 \Rightarrow (-.016, .116)$

$\sigma_{\hat{p}_2} = \sqrt{\dfrac{.09(.91)}{200}} = .020 \quad \hat{p}_2 \pm 3\sigma_{\hat{p}_2} \Rightarrow .09 \pm 3(.020) \Rightarrow .09 \pm .06 \Rightarrow (.03, .15)$

No. The interval $\hat{p}_1 \pm 3\sigma_{\hat{p}_1}$ contains 0.

e. $\sigma_{\hat{p}_1} = \sqrt{\dfrac{.95(.05)}{100}} = .022 \quad \hat{p}_1 \pm 3\sigma_{\hat{p}_1} \Rightarrow .95 \pm 3(.022) \Rightarrow .95 \pm .066 \Rightarrow (.884, 1.016)$

$\sigma_{\hat{p}_2} = \sqrt{\dfrac{.91(.09)}{200}} = .020 \quad \hat{p}_2 \pm 3\sigma_{\hat{p}_2} \Rightarrow .91 \pm 3(.020) \Rightarrow .91 \pm .06 \Rightarrow (.85, .97)$

No. The interval $\hat{p}_1 \pm 3\sigma_{\hat{p}_1}$ contains 1.

8.5 a. H_0: $(p_1 - p_2) = 0$
H_a: $(p_1 - p_2) \neq 0$

Will need to calculate the following:

$$\hat{p}_1 = \frac{320}{800} = .40 \quad \hat{p}_2 = \frac{400}{800} = .50 \quad \hat{p} = \frac{320 + 400}{800 + 800} = .45$$

The test statistic is $z = \dfrac{(\hat{p}_1 - \hat{p}_2) - 0}{\sqrt{\hat{p}\hat{q}\left(\dfrac{1}{n_1} + \dfrac{1}{n_2}\right)}} = \dfrac{(.40 - .50) - 0}{\sqrt{(.45)(.55)\left(\dfrac{1}{800} + \dfrac{1}{800}\right)}} = -4.02$

The rejection region requires $\alpha/2 = .05/2 = .025$ in each tail of the z-distribution. From Table III, Appendix A, $z_{.025} = 1.96$. The rejection region is $z < -1.96$ or $z > 1.96$.

Since the observed value of the test statistic falls in the rejection region ($-4.02 < -1.96$), H_0 is rejected. There is sufficient evidence to indicate that the proportions are unequal at $\alpha = .05$.

b. The problem is identical to part a until the rejection region. The rejection region requires $\alpha/2 = .01/2 = .005$ in each tail of the z-distribution. From Table III, Appendix A, $z_{.005} = 2.58$. The rejection region is $z < -2.58$ or $z > 2.58$.

Since the observed value of the test statistic falls in the rejection region ($-4.02 < -2.58$), H_0 is rejected. There is sufficient evidence to indicate that the proportions are unequal at $\alpha = .01$.

c. H_0: $p_1 - p_2 = 0$
H_a: $p_1 - p_2 < 0$

Test statistic as above: $z = -4.02$

The rejection region requires $\alpha = .01$ in the lower tail of the z-distribution. From Table III, Appendix A, $z_{.01} = 2.33$. The rejection region is $z < -2.33$.

Since the observed value of the test statistic falls in the rejection region ($-4.02 < -2.33$), H_0 is rejected. There is sufficient evidence to indicate that $p_1 < p_2$ at $\alpha = .01$.

d. $(\hat{p}_1 - \hat{p}_2) \pm z_{.05}\sqrt{\dfrac{\hat{p}_1\hat{q}_1}{n_1} + \dfrac{\hat{p}_2\hat{q}_2}{n_2}} \Rightarrow (.4 - .5) \pm (1.645)\sqrt{\dfrac{(.4)(.6)}{800} + \dfrac{(.5)(.5)}{800}}$
$\Rightarrow -.10 \pm .04 \Rightarrow (-.14, -.06)$

8.7 The sampling distribution $\hat{p}_1 - \hat{p}_2$ is approximately normal with:

$$\mu_{(\hat{p}_1-\hat{p}_2)} = p_1 - p_2 = .1 - .5 = -.4$$

$$\sigma_{(\hat{p}_1-\hat{p}_2)} = \sqrt{\frac{p_1q_1}{n_1} + \frac{p_2q_2}{n_2}} = \sqrt{\frac{(.1)(.9)}{100} + \frac{(.5)(.5)}{200}} = .046$$

8.9 Let p_1 = proportion of patients who suffered a first stroke who had antibodies and p_2 = proportion of patients hospitalized for other reasons who had antibodies.

Some preliminary calculations are:

$$\hat{p}_1 = \frac{x_1}{n_1} = \frac{25}{255} = .098; \ \hat{p}_2 = \frac{x_2}{n_2} = \frac{11}{257} = .043; \ \hat{p} = \frac{x_1 + x_2}{n_1 + n_2} = \frac{25 + 11}{255 + 257} = .070$$

To determine if the proportion of renegade antibodies in the blood is greater for patients who suffered a first stroke than for patients hospitalized for other reasons, we test:

H_0: $p_1 - p_2 = 0$
H_a: $p_1 - p_2 > 0$

The test statistic is $z = \dfrac{(\hat{p}_1 - \hat{p}_2) - 0}{\sqrt{\hat{p}\hat{q}\left(\dfrac{1}{n_1} + \dfrac{1}{n_2}\right)}} = \dfrac{(.098 - .043) - 0}{\sqrt{.07(.93)\left(\dfrac{1}{255} + \dfrac{1}{257}\right)}} = 2.44$

Since no α is given, we will use $\alpha = .05$. The rejection region requires $\alpha = .05$ in the upper tail of the z-distribution. From Table III, Appendix A, $z_{.05} = 1.645$. The rejection region is $z > 1.645$.

Since the observed value of the test statistic falls in the rejection region ($z = 2.44 > 1.645$), H_0 is rejected. There is sufficient evidence to indicate the proportion of first stroke patients who had antibodies is greater than the proportion of patients hospitalized for other reasons who had antibodies at $\alpha = .05$.

8.11 a. Let p_1 = proportion of female students who switched due to loss of interest in SME and p_2 = proportion of male students who switched due to lack of interest in SME.

Some preliminary calculations are:

$$\hat{p}_1 = \frac{x_1}{n_1} = \frac{74}{172} = .43; \ \hat{p}_2 = \frac{x_2}{n_2} = \frac{72}{163} = .44; \ \hat{p} = \frac{x_1 + x_2}{n_1 + n_2} = \frac{74 + 72}{172 + 163} = .436$$

To determine if the proportion of female students who switch due to lack of interest in SME differs from the proportion of males who switch due to a lack of interest, we test:

H_0: $p_1 - p_2 = 0$
H_a: $p_1 - p_2 \neq 0$

The test statistic is $z = \dfrac{(\hat{p}_1 - \hat{p}_2) - 0}{\sqrt{\hat{p}\hat{q}\left(\dfrac{1}{n_1} + \dfrac{1}{n_2}\right)}} = \dfrac{(.43 - .44) - 0}{\sqrt{.436(.564)\left(\dfrac{1}{172} + \dfrac{1}{163}\right)}} = -0.18$

The rejection region requires $\alpha/2 = .10/2 = .05$ in each tail of the z-distribution. From Table III, Appendix A, $z_{.05} = 1.645$. The rejection region is $z < -1.645$ or $z > 1.645$.

Since the observed value of the test statistic does not fall in the rejection region ($z = -0.18 \not< -1.645$), H_0 is not rejected. There is insufficient evidence to indicate the proportion of female students who switch due to lack of interest in SME differs from the proportion of males who switch due to a lack of interest in SME at $\alpha = .10$.

b. Let p_1 = proportion of female students who switched due to low grades in SME and p_2 = proportion of male students who switched due to low grades in SME.

Some preliminary calculations are:

$$\hat{p}_1 = \frac{x_1}{n_1} = \frac{33}{172} = .19; \ \hat{p}_2 = \frac{x_2}{n_2} = \frac{44}{163} = .27$$

For confidence coefficient .90, $\alpha = .10$ and $\alpha/2 = .10/2 = .05$. From Table III, Appendix A, $z_{.05} = 1.645$. The confidence interval is:

$$(\hat{p}_1 - \hat{p}_2) \pm z_{.05}\sqrt{\frac{\hat{p}_1\hat{q}_1}{n_1} + \frac{\hat{p}_2\hat{q}_2}{n_2}} \Rightarrow (.19 - .27) \pm 1.645\sqrt{\frac{.19(.81)}{172} + \frac{.27(.73)}{163}}$$
$$\Rightarrow -.08 \pm .075 \Rightarrow (-.155, -.005)$$

We are 90% confident that the difference between the proportions of female and male switchers who lost confidence due to low grades in SME is between $-.155$ and $-.005$. Since the interval does not include 0, there is evidence to indicate the proportion of female switchers due to low grades is less than the proportion of male switchers due to low grades.

8.13 a. Let p_1 = proportion of men managers who are married and p_2 = proportion of women managers who are married. For confidence coefficient .95, $\alpha = .05$ and $\alpha/2 = .025$. From Table III, Appendix A, $z_{.025} = 1.96$. The 95% confidence interval is:

$$(\hat{p}_1 - \hat{p}_2) \pm z_{.025}\sqrt{\frac{\hat{p}_1\hat{q}_1}{n_1} + \frac{\hat{p}_2\hat{q}_2}{n_2}} \Rightarrow (.86 - .45) \pm 1.96\sqrt{\frac{.86(.14)}{795} + \frac{.45(.55)}{223}}$$
$$\Rightarrow .41 \pm .070 \Rightarrow (.340, .480)$$

b. We are 95% confident that the difference in the proportion of married men and women managers from twenty Fortune 500 corporations is between .34 and .48.

Since 0 is not in the interval, there is evidence that the proportion of men managers who are married is greater than the proportion of women managers who are married.

8.15 a. To determine if a gender gap exists with females preferring the Democratic platform, we test:

H_0: $p_1 - p_2 = 0$
H_a: $p_1 - p_2 > 0$

where p_1 and p_2 are the proportions of males and females, respectively, who preferred the Republican platform.

Here, $\hat{p}_1 = \frac{81}{150} = .54 \quad \hat{p}_2 = \frac{70}{150} = .4667 \quad \hat{p} = \frac{81 + 70}{150 + 150} = .5033$

The test statistic is $z = \frac{(\hat{p}_1 - \hat{p}_2) - 0}{\sqrt{\hat{p}\hat{q}\left(\frac{1}{n_1} + \frac{1}{n_2}\right)}} = \frac{(.54 - .4667) - 0}{\sqrt{(.5033)(.4967)\left(\frac{1}{150} + \frac{1}{150}\right)}} = 1.27$

The rejection region requires $\alpha = .05$ in the upper tail of the z-distribution. From Table III, Appendix A, $z_{.05} = 1.645$. The rejection region is $z > 1.645$.

Since the observed value of the test statistic does not fall in the rejection region ($1.27 \not> 1.645$), H_0 is not rejected. There is insufficient evidence to indicate there is a gender gap at $\alpha = .05$.

b. To determine if women prefer the Democrats' responses to issues, we test:

H_0: $p = .5$
H_a: $p > .5$

where p = the proportion of females who prefer the Democrats' responses to issues.

$$\hat{p} = \frac{x}{n} = \frac{150 - 70}{150} = \frac{80}{150} = .5333$$

The test statistic is $z = \frac{\hat{p} - p_0}{\sigma_{\hat{p}}} = \frac{\hat{p} - p_0}{\sqrt{\frac{p_0 q_0}{n}}} = \frac{.5333 - .5}{\sqrt{\frac{.5(.5)}{150}}} = .82$

The rejection region requires $\alpha = .01$ in the upper tail of the z-distribution. From Table III, Appendix A, $z_{.01} = 2.33$. The rejection region is $z > 2.33$.

Since the observed value of the test statistic does not fall in the rejection region ($.82 \not> 2.33$), H_0 is not rejected. There is insufficient evidence to indicate women prefer the Democrats' responses to issues at $\alpha = .01$.

8.17 a. Let p_1 = proportion of cells that exhibit altered growth in the control group and p_2 = proportion of cells that exhibit altered growth in the E2F1 treated group.

Some preliminary calculations are:

$$\hat{p}_1 = \frac{x_1}{n_1} = \frac{15}{158} = .095; \ \hat{p}_2 = \frac{x_2}{n_2} = \frac{41}{92} = .446$$

For confidence coefficient .90, $\alpha = .10$ and $\alpha/2 = .10/2 = .05$. From Table III, Appendix A, $z_{.05} = 1.645$. The confidence interval is:

$$(\hat{p}_1 - \hat{p}_2) \pm z_{.05}\sqrt{\frac{\hat{p}_1\hat{q}_1}{n_1} + \frac{\hat{p}_2\hat{q}_2}{n_2}} \Rightarrow (.095 - .446) \pm 1.645\sqrt{\frac{.095(.905)}{158} + \frac{.446(.554)}{92}}$$
$$\Rightarrow -.351 \pm .093 \Rightarrow (-.444, -.258)$$

The difference in the percentages is (−44.4%, −25.8%).

We are 90% confident that the difference between the percentages of cells exhibiting altered growth between the control group and the group receiving E2F1 is between −44.4% and −25.8%.

b. Since 0 is not in the confidence interval, there is evidence to indicate the percentage of cells in the E2F1 group that exhibit altered growth is greater than the percentage of cells in the control group that exhibit altered growth. There is evidence to indicate the E2F1 transcription factor induces cell growth.

8.19 a. Let p_1 = proportion of employees in the Poor fitness category who show signs of stress and p_2 = proportion of employees in the Average fitness category who show signs of stress. To determine if a greater proportion of employees in the Poor fitness category show signs of stress than those in the Average fitness category, we test:

H_0: $p_1 - p_2 = 0$
H_a: $p_1 - p_2 > 0$

b. $$\hat{p} = \frac{n_1\hat{p}_1 + n_2\hat{p}_2}{n_1 + n_2} = \frac{242(.155) + 212(.133)}{242 + 212} = .145$$

The test statistic is $$z = \frac{(\hat{p}_1 - \hat{p}_2) - 0}{\sqrt{\hat{p}\hat{q}\left(\frac{1}{n_1} + \frac{1}{n_2}\right)}} = \frac{(.155 - .133) - 0}{\sqrt{.145(.855)\left(\frac{1}{242} + \frac{1}{212}\right)}} = .664$$

The rejection region requires $\alpha = .10$ in the upper tail of the z-distribution. From Table III, Appendix A, $z_{.10} = 1.28$. The rejection region is $z > 1.28$.

Since the observed value of the test statistic does not fall in the rejection region ($z = .664 > 1.28$), H_0 is not rejected. There is insufficient evidence to indicate that a greater proportion of employees in the Poor fitness category show signs of stress than those in the Average fitness category for $\alpha = .10$.

c. Let p_3 = proportion of employees in the Good fitness category who show signs of stress. To determine if a greater proportion of employees in the Poor fitness category show signs of stress than those in the Good fitness category, we test:

H_0: $p_1 - p_3 = 0$
H_a: $p_1 - p_3 > 0$

d. The test statistic is $z = 1.11$ and the p-value $= .1335$. Since the p-value is not small, there is no evidence to reject H_0. There is insufficient evidence to indicate that a greater proportion of employees in the Poor fitness category show signs of stress than those in the Good fitness category for $\alpha \leq .10$.

From these data, there is no evidence to indicate fitness level has a bearing on whether an individual shows signs of stress. This does not necessarily imply that fitness level has no bearing on whether an individual shows signs of stress. We just do not have enough evidence to say that it does have a bearing.

8.21 For confidence coefficient .90, $\alpha = 1 - .90 = .10$ and $\alpha/2 = .10/2 = .05$. From Table III, Appendix A, $z_{.05} = 1.645$. Since we have no prior information about the proportions, we use $p_1 = p_2 = .5$ to get a conservative estimate. For a width of .04, the bound is .02.

$$n_1 = n_2 = \frac{(z_{\alpha/2})^2(p_1q_1 + p_2q_2)}{B^2} = \frac{(1.645)^2(.5(1 - .5) + .5(1 - .5))}{.02^2} = 3382.5 \approx 3383$$

8.23 For probability .95, $\alpha = 1 - .95 = .05$ and $\alpha/2 = .05/2 = .025$. From Table III, Appendix A, $z_{.025} = 1.96$. Since we have no prior information about the proportions, we use $p_1 = p_2 = .5$ to get a conservative estimate.

$$n_1 = n_2 = \frac{(z_{\alpha/2})^2(p_1q_1 + p_2q_2)}{B^2} = \frac{(1.96)^2(.5(1 - .5) + .5(1 - .5))}{.02^2} = \frac{1.9208}{.0004} = 4802$$

8.25 a. For confidence coefficient .95, $\alpha = .05$ and $\alpha/2 = .025$. From Table III, Appendix A, $z_{.025} = 1.96$.

$$n_1 = n_2 = \frac{(z_{\alpha/2})^2(p_1q_1 + p_2q_2)}{B^2} = \frac{1.96^2((.155)(.845) + (.108)(.892))}{.04^2} = 545.8 \approx 546$$

b. Let p_1 = proportion of employees in the Poor fitness category who show signs of stress and p_2 = proportion of employees in the Good fitness category who show signs of stress. To determine if a greater proportion of employees in the Poor fitness category show signs of stress than those in the Good fitness category, we test:

H_0: $p_1 - p_2 = 0$
H_a: $p_1 - p_2 > 0$

$$\hat{p} = \frac{n_1\hat{p}_1 + n_2\hat{p}_2}{n_1 + n_2} = \frac{546(.155) + 546(.108)}{546 + 546} = .1315$$

The test statistic is $z = \dfrac{(\hat{p}_1 - \hat{p}_2) - 0}{\sqrt{\hat{p}\hat{q}\left(\dfrac{1}{n_1} + \dfrac{1}{n_2}\right)}} = \dfrac{(.155 - .108) - 0}{\sqrt{.1315(.8685)\left(\dfrac{1}{546} + \dfrac{1}{546}\right)}} = 2.30$

The p-value $= P(z \geq 2.30) = .5 - .4893 = .0107$. Since this p-value is so small, there is evidence to reject H_0. There is sufficient evidence to indicate that a greater proportion of employees in the Poor fitness category show signs of stress than those in the Good fitness category for $\alpha > .0107$.

8.27 a. The rejection region requires $\alpha = .05$ in the upper tail of the χ^2 distribution with $df = k - 1 = 3 - 1 = 2$. From Table XI, Appendix A, $\chi^2_{.05} = 5.99147$. The rejection region is $\chi^2 > 5.99147$.

b. The rejection region requires $\alpha = .10$ in the upper tail of the χ^2 distribution with $df = k - 1 = 5 - 1 = 4$. From Table XI, Appendix A, $\chi^2_{.10} = 7.77944$. The rejection region is $\chi^2 > 7.77944$.

c. The rejection region requires $\alpha = .01$ in the upper tail of the χ^2 distribution with $df = k - 1 = 4 - 1 = 3$. From Table XI, Appendix A, $\chi^2_{.01} = 11.3449$. The rejection region is $\chi^2 > 11.3449$.

8.29 The sample size n will be large enough so that for every cell the expected cell count $E(n_i)$ will be equal to 5 or more.

8.31 Some preliminary calculations are:

If the probabilities are the same, $p_{1,0} = p_{2,0} = p_{3,0} = p_{4,0} = .25$

$$E(n_1) = np_{1,0} = 205(.25) = 51.25$$
$$E(n_2) = E(n_3) = E(n_4) = 205(.25) = 51.25$$

a. To determine if the multinomial probabilities differ, we test:

H_0: $p_1 = p_2 = p_3 = p_4 = .25$
H_a: At least one of the probabilities differs from .25

The test statistic is $\chi^2 = \sum \frac{[n_i - E(n_i)]^2}{E(n_i)}$

$$= \frac{(43 - 51.25)^2}{51.25} + \frac{(56 - 51.25)^2}{51.25} + \frac{(59 - 51.25)^2}{51.25} + \frac{(47 - 51.25)^2}{51.25} = 3.293$$

The rejection region requires $\alpha = .05$ in the upper tail of the χ^2 distribution with $df = k - 1 = 4 - 1 = 3$. From Table XI, Appendix A, $\chi^2_{.05} = 7.81473$. The rejection region is $\chi^2 > 7.81473$.

Since the observed value of the test statistic does not fall in the rejection region ($\chi^2 = 3.293 \not> 7.81473$), H_0 is not rejected. There is insufficient evidence to indicate the multinomial probabilities differ at $\alpha = .05$.

b. The Type I error is concluding the multinomial probabilities differ when, in fact, they do not.

The Type II error is concluding the multinomial probabilities are equal, when, in fact, they are not.

8.33 To determine if the true percentages of ADEs in the five "cause" categories are different, we test:

H_0: $p_1 = p_2 = p_3 = p_4 = p_5 = .2$
H_a: At least one p_i differs from .2, $i = 1, 2, 3, 4, 5$

The test statistic is $\chi^2 = 16$ (from printout).

The p-value of the test is $p = 3.01916\text{E-}3 = .00301916$.

Since the p-value is less than α ($p = .00301916 < .10$), H_0 is rejected. There is sufficient evidence to indicate that at least one percentage of ADEs in the five "cause" categories is different at $\alpha = .10$.

8.35 a. To determine if a difference in the preference of entrepreneurs for the cars, we test:

H_0: $p_1 = p_2 = p_3 = 1/3$
H_a: At least one of the multinomial probabilities does not equal 1/3

Where p_1 = proportion who prefer U.S. cars
p_2 = proportion who prefer European cars
p_3 = proportion who prefer Japanese cars

$$E(n_1) = np_{1,0} = 100\left(\frac{1}{3}\right) = 33\frac{1}{3}$$

$$E(n_2) = np_{2,0} = 100\left(\frac{1}{3}\right) = 33\frac{1}{3}$$

$$E(n_3) = np_{3,0} = 100\left(\frac{1}{3}\right) = 33\frac{1}{3}$$

The test statistic is $\chi^2 = \sum \frac{[n_i - E(n_i)]^2}{E(n_i)}$

$$= \frac{\left(45 - 33\frac{1}{3}\right)^2}{33\frac{1}{3}} + \frac{\left(46 - 33\frac{1}{3}\right)^2}{33\frac{1}{3}} + \frac{\left(9 - 33\frac{1}{3}\right)^2}{33\frac{1}{3}} = 26.66$$

The rejection region requires $\alpha = .05$ in the upper tail of the χ^2 distribution with df = $k - 1 = 3 - 1 = 2$. From Table XI, Appendix A, $\chi^2_{.05} = 5.99147$. The rejection region is $\chi^2 > 5.99147$.

Since the observed value of the test statistic falls in the rejection region ($\chi^2 = 26.66 > 5.99147$), H_0 is rejected. There is sufficient evidence to indicate a difference in the preference of entrepreneurs for the cars at $\alpha = .05$.

b. To determine if there is a difference in the preference of entrepreneurs for domestic versus foreign cars, we test:

H_0: $p_1 = p_2 = .5$
H_a: At least one probability differs from .5

Where p_1 = proportion who prefer U.S. cars
p_2 = proportion who prefer foreign cars

$E(n_1) = np_{1,0} = 100(.5) = 50$
$E(n_2) = np_{2,0} = 100(.5) = 50$

The test statistic is $\chi^2 = \sum \frac{\sum[n_i - E(n_i)]^2}{E(n_i)} = \frac{(45 - 50)^2}{50} + \frac{(55 - 50)^2}{50} = 1$

The rejection region requires $\alpha = .05$ in the upper tail of the χ^2 distribution with df = $k - 1 = 2 - 1 = 1$. From Table XI, Appendix A, $\chi^2_{.05} = 3.84146$. The rejection region is $\chi^2 > 3.84146$.

Since the observed value of the test statistic does not fall in the rejection region ($\chi^2 = 1 \not> 3.84146$), H_0 is not rejected. There is insufficient evidence to indicate a difference in the preference of entrepreneurs for domestic and foreign cars at $\alpha = .05$.

c. We must assume the sample size n is large enough so that the expected cell count, $E(n_i)$, will be equal to 5 or more for every cell.

8.37 Some preliminary calculations:

$$E(n_1) = np_{1,0} = 90(1/3) = 30$$
$$E(n_2) = np_{2,0} = 90(1/3) = 30$$
$$E(n_3) = np_{3,0} = 90(1/3) = 30$$

To determine if the true proportions of subjects that fall in the three disk orientation categories differ, we test:

H_0: $p_1 = p_2 = p_3 = 1/3$
H_a: At least one p_i differs from 1/3, $i = 1, 2, 3$

The test statistic is $\chi^2 = \sum \frac{[n_i - E(n_i)]^2}{E(n_i)} = \frac{(58 - 30)^2}{30} + \frac{(15 - 30)^2}{30} + \frac{(17 - 30)^2}{30} = 39.267$

The rejection region requires $\alpha = .05$ in the upper tail of the χ^2 distribution with df = $k - 1 = 3 - 1 = 2$. From Table XI, Appendix A, $\chi^2_{.05} = 5.99147$. The rejection region is $\chi^2 > 5.99147$.

Since the observed value of the test statistic falls in the rejection region ($\chi^2 = 39.267 > 5.99147$), H_0 is rejected. There is sufficient evidence to indicate that at least one of the true proportions of subjects that fall in the three disk orientation categories differs at $\alpha = .05$.

8.39 Some preliminary calculations:

$$E(n_1) = np_{1,0} = 2{,}588(.05) = 129.4$$
$$E(n_2) = np_{2,0} = 2{,}588(.05) = 129.4$$
$$E(n_3) = np_{3,0} = 2{,}588(.90) = 2{,}392.2$$

To determine if the theory on hotspots/coldspots holds, we test:

H_0: $p_1 = p_2 = .05$, $p_3 = .90$
H_a: At least one p_i differs from its hypothesized value

The test statistic is $\chi^2 = \sum \frac{[n_i - E(n_i)]^2}{E(n_i)}$

$$= \frac{(123 - 129.4)^2}{129.4} + \frac{(147 - 129.4)^2}{129.4} + \frac{(2318 - 2329.2)^2}{2329.2} = 2.764$$

The rejection region requires $\alpha = .01$ in the upper tail of the χ^2 distribution with df = $k - 1 = 3 - 1 = 2$. From Table XI, Appendix A, $\chi^2_{.01} = 9.21034$. The rejection region is $\chi^2 > 9.21034$.

Since the observed value of the test statistic does not fall in the rejection region ($\chi^2 = 2.764 \not> 9.21034$), H_0 is not rejected. There is insufficient evidence to indicate that at least one of the true proportions of hotspots/coldspots differs from the hypothesized value at $\alpha = .01$.

8.41 a. H_0: The row and column classifications are independent
H_a: The row and column classifications are dependent

b. The test statistic is $\chi^2 = \sum\sum \frac{\left[n_{ij} - \hat{E}(n_{ij})\right]^2}{\hat{E}(n_{ij})}$

The rejection region requires $\alpha = .01$ in the upper tail of the χ^2 distribution with df $= (r - 1)(c - 1) = (2 - 1)(3 - 1) = 2$. From Table XI, Appendix A, $\chi^2_{.01} = 9.21034$. The rejection region is $\chi^2 > 9.21034$.

c. The expected cell counts are:

$$\hat{E}(n_{11}) = \frac{r_1c_1}{n} = \frac{96(25)}{167} = 14.37 \qquad E(n_{21}) = \frac{r_2c_1}{n} = \frac{71(25)}{167} = 10.63$$

$$\hat{E}(n_{12}) = \frac{r_1c_2}{n} = \frac{96(64)}{167} = 36.79 \qquad \hat{E}(n_{22}) = \frac{r_2c_2}{n} = \frac{71(64)}{167} = 27.21$$

$$\hat{E}(n_{13}) = \frac{r_1c_3}{n} = \frac{96(78)}{167} = 44.84 \qquad \hat{E}(n_{23}) = \frac{r_2c_3}{n} = \frac{71(78)}{167} = 33.16$$

d. The test statistic is $\chi^2 = \sum\sum \frac{\left[n_{ij} - \hat{E}(n_{ij})\right]^2}{\hat{E}(n_{ij})}$

$$= \frac{(9 - 14.37)^2}{14.37} + \frac{(34 - 36.79)^2}{36.79} + \frac{(53 - 44.84)^2}{44.84} + \frac{(16 - 10.63)^2}{10.63} + \frac{(30 - 27.21)^2}{27.21} + \frac{(25 - 33.16)^2}{33.16} = 8.71$$

Since the observed value of the test statistic does not fall in the rejection region ($\chi^2 = 8.71 \not> 9.21034$), H_0 is not rejected. There is insufficient evidence to indicate the row and column classifications are dependent at $\alpha = .01$.

8.43 Some preliminary calculations are:

$$\hat{E}(n_{11}) = \frac{r_1c_1}{n} = \frac{154(134)}{439} = 47.007 \qquad \hat{E}(n_{21}) = \frac{186(134)}{439} = 56.774$$

$$\hat{E}(n_{12}) = \frac{154(163)}{439} = 57.180 \qquad \hat{E}(n_{22}) = \frac{186(163)}{439} = 69.062$$

$$\hat{E}(n_{13}) = \frac{154(142)}{439} = 49.813 \qquad \hat{E}(n_{23}) = \frac{186(142)}{439} = 60.164$$

$$\hat{E}(n_{31}) = \frac{99(134)}{439} = 30.219 \qquad \hat{E}(n_{33}) = \frac{99(142)}{439} = 32.023$$

$$\hat{E}(n_{32}) = \frac{99(163)}{439} = 36.759$$

To determine if the row and column classifications are dependent, we test:

H_0: The row and column classifications are independent
H_a: The row and column classifications are dependent

The test statistic is $\chi^2 = \sum\sum \frac{[n_{ij} - \hat{E}(n_{ij})]^2}{\hat{E}(n_{ij})}$

$$= \frac{(40 - 47.007)^2}{47.007} + \frac{(72 - 57.180)^2}{57.180} + \frac{(42 - 49.813)^2}{49.813} + \frac{(63 - 56.774)^2}{56.774}$$
$$+ \frac{(53 - 69.062)^2}{69.062} + \frac{(70 - 60.164)^2}{60.164} + \frac{(31 - 30.219)^2}{30.219}$$
$$+ \frac{(38 - 36.759)^2}{36.759} + \frac{(30 - 32.023)^2}{32.023} = 12.36$$

The rejection region requires $\alpha = .05$ in the upper tail of the χ^2 distribution with df = $(r - 1)(c - 1) = (3 - 1)(3 - 1) = 4$. From Table XI, Appendix A, $\chi^2_{.05} = 9.48773$. The rejection region is $\chi^2 > 9.48773$.

Since the observed value of the test statistic falls in the rejection region ($\chi^2 = 12.36 > 9.48773$), H_0 is rejected. There is sufficient evidence to indicate the row and column classification are dependent at $\alpha = .05$.

8.45 a. The sample proportion of injured Hispanic children who were not wearing seatbelts during the accident is:

$$\hat{p} = 283/314 = .901$$

b. The sample proportion of injured non-Hispanic white children who were not wearing seatbelts during the accident is:

$$\hat{p} = 330/478 = .690$$

c. Since the proportion of injured Hispanic children who were not wearing seatbelts during the accident (.901) is .211 higher than the proportion of injured non-Hispanic white children who were not wearing seatbelts during the accident (.690), the proportions probably differ.

d. Some preliminary calculations are:

$$\hat{E}(n_{11}) = \frac{r_1c_1}{n} = \frac{179(314)}{792} = 70.97 \qquad \hat{E}(n_{12}) = \frac{r_1c_2}{n} = \frac{179(478)}{792} = 108.03$$
$$\hat{E}(n_{21}) = \frac{r_2c_1}{n} = \frac{613(314)}{792} = 243.03 \qquad \hat{E}(n_{22}) = \frac{r_2c_2}{n} = \frac{613(478)}{792} = 369.97$$

To determine whether seatbelt usage in motor vehicle accidents depends on ethnic status in the San Diego County Regionalized Trauma System, we test:

H_0: Seatbelt usage in motor vehicle accidents and ethnic status in the San Diego County Regionalized Trauma System are independent
H_a: Seatbelt usage in motor vehicle accidents and ethnic status in the San Diego County Regionalized Trauma System are dependent

The test statistic is $\chi^2 = \sum\sum \frac{\left[n_{ij} - \hat{E}(n_{ij})\right]^2}{\hat{E}(n_{ij})^2}$

$$= \frac{(31 - 70.97)^2}{70.97} + \frac{(148 - 108.03)^2}{108.03} + \frac{(283 - 243.03)^2}{243.03} + \frac{(330 - 369.97)^2}{369.97} = 48.191$$

The rejection region requires $\alpha = .01$ in the upper tail of the χ^2 distribution with df = $(r - 1)(c - 1) = (2 - 1)(2 - 1) = 1$. From Table XI, Appendix A, $\chi^2_{.01} = 6.63490$. The rejection region is $\chi^2 > 6.63490$.

Since the observed value of the test statistic falls in the rejection region ($\chi^2 = 48.191 > 6.63490$), H_0 is rejected. There is sufficient evidence to indicate seatbelt usage in motor vehicle accidents depends on ethnic status in the San Diego County Regionalized Trauma System at $\alpha = .01$.

e. For confidence coefficient .99, $\alpha = .01$ and $\alpha/2 = .01/2 = .005$. From Table III, Appendix A, $z_{.005} = 2.58$. The confidence interval is:

$$(\hat{p}_1 - \hat{p}_2) \pm z_{.005}\sqrt{\frac{\hat{p}_1\hat{q}_1}{n_1} + \frac{\hat{p}_2\hat{q}_2}{n_2}} \Rightarrow (.901 - .690) \pm 2.58\sqrt{\frac{.901(.099)}{314} + \frac{.690(.310)}{478}}$$
$$\Rightarrow .211 \pm .070 \Rightarrow (.141, .281)$$

We are 99% confident that the difference in the proportion of injured Hispanic children who were not wearing seatbelts and the proportion of injured non-Hispanic white children who were not wearing seatbelts is between .141 and .281.

8.47 a. Some preliminary calculations are:

$$\hat{E}(n_{11}) = \frac{r_1c_1}{n} = \frac{136(150)}{300} = 68 \qquad \hat{E}(n_{21}) = \frac{r_2c_1}{n} = \frac{164(150)}{300} = 82$$

$$\hat{E}(n_{12}) = \frac{r_1c_2}{n} = \frac{136(150)}{300} = 68 \qquad \hat{E}(n_{22}) = \frac{r_2c_2}{n} = \frac{164(150)}{300} = 82$$

To determine whether audience gender and product identification are dependent factors for male spokespersons, we test:

H_0: Audience gender and product identification are independent factors
H_a: Audience gender and product identification are dependent factors

The test statistic is $\chi^2 = \sum\sum \frac{\left[n_{ij} - \hat{E}(n_{ij})\right]^2}{\hat{E}(n_{ij})^2}$

$$= \frac{(95 - 68)^3}{68} + \frac{(41 - 68)^2}{68} + \frac{(55 - 82)^2}{82} + \frac{(109 - 82)^2}{82} = 39.22$$

The rejection region requires $\alpha = .05$ in the upper tail of the χ^2 distribution with df = $(r - 1)(c - 1) = (2 - 1)(2 - 1) = 1$. From Table XI, Appendix A, $\chi^2_{.05} = 3.84146$. The rejection region is $\chi^2 > 3.84146$.

Since the observed value of the test statistic falls in the rejection region ($\chi^2 = 39.22 > 3.84146$), H_0 is rejected. There is sufficient evidence to indicate audience gender and product identification are dependent factors for $\alpha = .05$.

b. Some preliminary calculations are:

$$\hat{E}(n_{11}) = \frac{r_1 c_1}{n} = \frac{108(150)}{300} = 54 \qquad \hat{E}(n_{21}) = \frac{r_2 c_1}{n} = \frac{192(150)}{300} = 96$$

$$\hat{E}(n_{12}) = \frac{r_1 c_2}{n} = \frac{108(150)}{300} = 54 \qquad \hat{E}(n_{22}) = \frac{r_2 c_2}{n} = \frac{192(150)}{300} = 96$$

To determine whether audience gender and product identification are dependent factors for female spokespersons, we test:

H_0: Audience gender and product identification are independent factors
H_a: Audience gender and product identification are dependent factors

The test statistic is $\chi^2 = \sum\sum \frac{\left[n_{ij} - \hat{E}(n_{ij})\right]^2}{\hat{E}(n_{ij})}$

$$= \frac{(47 - 54)^2}{54} + \frac{(61 - 54)^2}{54} + \frac{(103 - 96)^2}{96} + \frac{(89 - 96)^2}{96} = 2.84$$

The rejection region requires $\alpha = .05$ in the upper tail of the χ^2 distribution with df = $(r - 1)(c - 1) = (2 - 1)(2 - 1) = 1$. From Table XI, Appendix A, $\chi^2_{.05} = 3.84146$. The rejection region is $\chi^2 > 3.84146$.

Since the observed value of the test statistic does not fall in the rejection region ($\chi^2 = 2.84 \not> 3.84146$), H_0 is not rejected. There is insufficient evidence to indicate audience gender and product identification are dependent factors for $\alpha = .05$.

c. When a male spokesperson is used in an advertisement, audience gender and product identification are dependent. Males tended to identify the product more frequently than females.

When a female spokesperson is used in an advertisement, there is no evidence that audience gender and product identification are dependent. Males and females tend to identify the product at the same rate.

8.49 First, we check to see if the assumption about the expected cells is met. From the table, there are two expected cell counts that are less than 5. Thus, the results of the test are suspect.

To determine if the number of kills is related to whether the trial was stopped or not, we test:

H_0: Number of kills and whether the trial was stopped or not are independent
H_a: Number of kills and whether the trial was stopped or not are dependent

The test statistic is $\chi^2 = 2.17067$ (from the printout).

The p-value of the test is .70440. Since this p-value is so large, H_0 is not rejected. There is insufficient evidence to indicate that the number of kills is related to whether the trial was stopped or not at $\alpha \leq .10$.

8.51 Some preliminary calculations are:

$$\hat{E}(n_{11}) = \frac{r_1c_1}{n} = \frac{419(380)}{703} = 226.49 \qquad \hat{E}(n_{21}) = \frac{r_2c_1}{n} = \frac{149(380)}{703} = 80.54$$

$$\hat{E}(n_{12}) = \frac{r_1c_2}{n} = \frac{419(323)}{703} = 192.51 \qquad \hat{E}(n_{22}) = \frac{r_2c_2}{n} = \frac{149(323)}{703} = 68.46$$

$$\hat{E}(n_{31}) = \frac{r_3c_1}{n} = \frac{87(380)}{703} = 47.03 \qquad \hat{E}(n_{41}) = \frac{r_4c_1}{n} = \frac{48(380)}{703} = 25.95$$

$$\hat{E}(n_{32}) = \frac{r_3c_2}{n} = \frac{87(323)}{703} = 39.97 \qquad \hat{E}(n_{42}) = \frac{r_4c_2}{n} = \frac{48(323)}{703} = 22.05$$

To determine whether retirement status of a traveler and the duration of a typical trip are dependent, we test:

H_0: Retirement status of a traveler and the duration of a typical trip are independent
H_a: Retirement status of a traveler and the duration of a typical trip are dependent

The test statistic is $\chi^2 = \sum\sum \frac{\left[n_{ij} - \hat{E}(n_{ij})\right]^2}{\hat{E}(n_{ij})}$

$$= \frac{(247 - 226.49)^2}{226.49} + \frac{(172 - 192.51)^2}{192.51} + \frac{(82 - 80.54)^2}{80.54} + \frac{(67 - 68.46)^2}{68.46}$$

$$+ \frac{(35 - 47.03)^2}{47.03} + \frac{(52 - 39.97)^2}{39.97} + \frac{(16 - 25.95)^2}{25.95} + \frac{(32 - 22.05)^2}{22.05} = 19.10$$

The rejection region requires $\alpha = .05$ in the upper tail of the χ^2 distribution with df = $(r - 1)(c - 1) = (4 - 1)(2 - 1) = 3$. From Table XI, Appendix A, $\chi^2_{.05} = 7.81473$. The rejection region is $\chi^2 > 7.81473$.

Since the observed value of the test statistic falls in the rejection region ($\chi^2 = 19.10 > 7.81473$), H_0 is rejected. There is sufficient evidence to indicate retirement status of a traveler and the duration of a typical trip are dependent for $\alpha = .05$.

8.53 a. No. In a multinomial experiment, each trial results in one of $k = 8$ possible outcomes. In this experiment, each trial cannot result in one of all of the eight possible outcomes. The people are assigned to a particular diet before the experiment. Once the people are assigned to a particular diet, there are only two possible outcomes, cancer tumors or no cancer tumors.

b. The expected cell counts are:

$$\hat{E}(n_{11}) = \frac{r_1c_1}{n} = \frac{80(30)}{120} = 20 \qquad \hat{E}(n_{21}) = \frac{r_2c_1}{n} = \frac{40(30)}{120} = 10$$

$$\hat{E}(n_{12}) = \frac{r_1c_2}{n} = \frac{80(30)}{120} = 20 \qquad \hat{E}(n_{22}) = \frac{r_2c_2}{n} = \frac{40(30)}{120} = 10$$

$$\hat{E}(n_{13}) = \frac{r_1c_3}{n} = \frac{80(30)}{120} = 20 \qquad \hat{E}(n_{23}) = \frac{r_2c_3}{n} = \frac{40(30)}{120} = 10$$

$$\hat{E}(n_{14}) = \frac{r_1c_4}{n} = \frac{80(30)}{120} = 20 \qquad \hat{E}(n_{24}) = \frac{r_2c_4}{n} = \frac{40(30)}{120} = 10$$

c. The test statistic is $\chi^2 = \sum\sum \frac{\left[n_{ij} - \hat{E}(n_{ij})\right]^2}{\hat{E}(n_{ij})}$

$$= \frac{(27 - 20)^2}{20} + \frac{(20 - 20)^2}{20} + \frac{(19 - 20)^2}{20} + \frac{(14 - 20)^2}{20} + \frac{(3 - 10)^2}{10} + \frac{(10 - 10)^2}{10} + \frac{(11 - 10)^2}{10} + \frac{(16 - 10)^2}{10} = 12.9$$

d. To determine if diet and presence/absence of cancer are independent, we test:

H_0: Diet and presence/absence of cancer are independent
H_a: Diet and presence/absence of cancer are dependent

The test statistic is $\chi^2 = 12.9$ (from part c).

The rejection region requires $\alpha = .05$ in the upper tail of the χ^2 distribution with df = $(r - 1)(c - 1) = (2 - 1)(4 - 1) = 3$. From Table XI, Appendix A, $\chi^2_{.05} = 7.81473$. The rejection region is $\chi^2 > 7.81473$.

Since the observed value of the test statistic falls in the rejection region ($\chi^2 = 12.9 > 7.81473$), H_0 is rejected. There is sufficient evidence to indicate that diet and presence/absence of cancer are dependent for $\alpha = .05$.

e. $\hat{p}_1 = 27/30 = .9$; $\hat{p}_2 = 20/30 = .667$

For confidence coefficient .95, $\alpha = .05$ and $\alpha/2 = .05/2 = .025$. From Table III, Appendix A, $z_{.025} = 1.96$. The confidence interval is:

$$(\hat{p}_1 - \hat{p}_2) \pm z_{.005}\sqrt{\frac{\hat{p}_1\hat{q}_1}{n_1} + \frac{\hat{p}_2\hat{q}_2}{n_2}} \Rightarrow (.900 - .667) \pm 1.96\sqrt{\frac{.900(.100)}{30} + \frac{.667(.333)}{30}}$$
$$\Rightarrow .233 \pm .200 \Rightarrow (.033, .433)$$

We are 95% confident that the difference in the percentage of rats on a high fat/no fiber diet with cancer and the percentage of rats on a high fat/fiber diet with cancer is between 3.3% and 43.3%.

8.55 Some preliminary calculations are:

$$\hat{p}_1 = \frac{x_1}{n_1} = \frac{110}{200} = .55;\ \hat{p}_2 = \frac{x_2}{n_2} = \frac{130}{200} = .65;\ \hat{p} = \frac{x_1 + x_2}{n_1 + n_2} = \frac{110 + 130}{200 + 200} = \frac{240}{400} = .60$$

a. H_0: $(p_1 - p_2) = 0$
H_a: $(p_1 - p_2) < 0$

The test statistic is $z = \frac{(\hat{p}_1 - \hat{p}_2) - 0}{\sqrt{\hat{p}\hat{q}\left(\frac{1}{n_1} + \frac{1}{n_2}\right)}} = \frac{(.55 - .65) - 0}{\sqrt{.6(1 - .6)\left(\frac{1}{200} + \frac{1}{200}\right)}} = \frac{-.10}{.049} = -2.04$

The rejection region requires $\alpha = .10$ in the lower tail of the z-distribution. From Table III, Appendix A, $z_{.10} = 1.28$. The rejection region is $z < -1.28$.

Since the observed value of the test statistic falls in the rejection region ($z = -2.04 < -1.28$), H_0 is rejected. There is sufficient evidence to conclude $(p_1 - p_2 < 0)$ at $\alpha = .10$.

b. For confidence coefficient .95, $\alpha = 1 - .95 = .05$ and $\alpha/2 = .05/2 = .025$. From Table III, Appendix A, $z_{.025} = 1.96$. The 95% confidence interval for $(p_1 - p_2)$ is approximately:

$$(\hat{p}_1 - \hat{p}_2) \pm z_{\alpha/2}\sqrt{\frac{\hat{p}_1\hat{q}_1}{n_1} + \frac{\hat{p}_2\hat{q}_2}{n_2}}$$

$$\Rightarrow (.55 - .65) \pm 1.96\sqrt{\frac{.55(1 - .55)}{200} + \frac{.65(1 - .65)}{200}}$$

$$\Rightarrow -.10 \pm .096 \Rightarrow (-.196, -.004)$$

c. From part **b**, $z_{.025} = 1.96$. Using the information from our samples, we can use $p_1 = .55$ and $p_2 = .65$. For a width of .01, the bound is .005.

$$n_1 = n_2 = \frac{(z_{\alpha/2})^2(p_1q_1 + p_2q_2)}{B^2} = \frac{(1.96)^2(.55(1 - .55) + .65(1 - .65))}{.005^2} = \frac{1.82476}{.000025}$$

$$= 72990.4 \approx 72{,}991$$

8.57 a. Some preliminary calculations are:

If all the categories are equally likely,

$$p_{1,0} = p_{2,0} = p_{3,0} = p_{4,0} = p_{5,0} = .2$$

$$E(n_1) = E(n_2) = E(n_3) = E(n_4) = E(n_5) = np_{1,0} = 150(.2) = 30$$

To determine if the categories are not equally likely, we test:

H_0: $p_1 = p_2 = p_3 = p_4 = p_5 = .2$
H_a: At least one probability is different from .2

The test statistic is $\chi^2 = \sum \frac{[n_i - E(n_i)]^2}{E(n_i)}$

$$= \frac{(28 - 30)^2}{30} + \frac{(35 - 30)^2}{30} + \frac{(33 - 30)^2}{30} + \frac{(25 - 30)^2}{30} + \frac{(29 - 30)^2}{30} = 2.133$$

The rejection region requires $\alpha = .10$ in the upper tail of the χ^2 distribution with df = $k - 1 = 5 - 1 = 4$. From Table XI, Appendix A, $\chi^2_{.10} = 7.77944$. The rejection region is $\chi^2 > 7.77944$.

Since the observed value of the test statistic does not fall in the rejection region ($\chi^2 = 2.133 \not> 7.77944$), H_0 is not rejected. There is insufficient evidence to indicate the categories are not equally likely at $\alpha = .10$.

b. $\hat{p}_2 = \frac{35}{150} = .233$

For confidence coefficient .90, $\alpha = .10$ and $\alpha/2 = .05$. From Table III, Appendix A, $z_{.05} = 1.645$. The confidence interval is:

$$\hat{p}_2 \pm z_{.05}\sqrt{\frac{\hat{p}_2\hat{q}_2}{n_2}} \Rightarrow .233 \pm 1.645\sqrt{\frac{.233(.767)}{150}} \Rightarrow .233 \pm .057 \Rightarrow (.176, .290)$$

8.59 a. Let p_1 = proportion of dead insects in the room sprayed with type A and p_2 = proportion of dead insects in the room sprayed with type B. Since we wish to determine if spray A is more effective than spray B, we test:

$$\hat{p}_1 = \frac{120}{200} = .6,\ \hat{p}_2 = \frac{90}{200} = .45$$

$$\hat{p} = \frac{x_1 + x_2}{n_1 + n_2} = \frac{120 + 90}{200 + 200} = \frac{210}{400} = .525$$

H_0: $(p_1 - p_2) = 0$
H_a: $(p_1 - p_2) > 0$

The test statistic is $z = \dfrac{(\hat{p}_1 - \hat{p}_2) - 0}{\sqrt{\hat{p}\hat{q}\left(\frac{1}{n_1} + \frac{1}{n_2}\right)}}$

$$z = \frac{(.6 - .45) - 0}{\sqrt{.525(1 - .525)\left(\frac{1}{200} + \frac{1}{200}\right)}} = \frac{.15}{.0499} = 3.00$$

The rejection region requires $\alpha = .05$ in the upper tail of the z-distribution. From Table III, Appendix A, $z_{.05} = 1.645$. The rejection region is $z > 1.645$.

Since the observed value of the test statistic falls in the rejection region ($z = 3.00 > 1.645$), H_0 is rejected. There is sufficient evidence to indicate spray A is more effective than spray B in controlling the insects ($p_1 - p_2 > 0$) at $\alpha = .05$.

b. For confidence coefficient .90, $\alpha = 1 - .90 = .10$ and $\alpha/2 = .10/2 = .05$. From Table III, Appendix A, $z_{.05} = 1.645$. The 90% confidence interval for $(p_1 - p_2)$ is:

$$(\hat{p}_1 - \hat{p}_2) \pm z_{\alpha/2}\sqrt{\frac{\hat{p}_1\hat{q}_1}{n_1} + \frac{\hat{p}_2\hat{q}_2}{n_2}}$$

$$\Rightarrow (.6 - .45) \pm 1.645\sqrt{\frac{.6(1 - .6)}{200} + \frac{.45(1 - .45)}{200}}$$

$$\Rightarrow .15 \pm .081 \Rightarrow (.069, .231)$$

We estimate the difference in the rates of kill for the two sprays to fall between .069 and .231. In other words, we estimate that p_1, the rate of kill for spray A, is larger than p_2, the rate of kill for spray B by .069 to .231.

8.61 a. To determine if the production of earthquake-insured residents differ for the four samples, we test:

H_0: Insurance status and county are independent
H_a: Insurance status and county are dependent

b. First we construct the contingency table:

	Counties				
Insurance Status	Contra Costa	Santa Clara	Los Angeles	San Bernardino	Total
Insured	117	222	133	109	581
Not insured	404	334	204	263	1205
Total	521	556	337	372	1786

Some preliminary calculations are:

$$\hat{E}(n_{11}) = \frac{r_1c_1}{n} = \frac{581(521)}{1786} = 169.49 \qquad \hat{E}(n_{21}) = \frac{r_2c_1}{n} = \frac{1205(521)}{1786} = 351.51$$

$$\hat{E}(n_{12}) = \frac{r_1c_2}{n} = \frac{581(556)}{1786} = 180.87 \qquad \hat{E}(n_{22}) = \frac{r_2c_2}{n} = \frac{1205(556)}{1786} = 375.13$$

$$\hat{E}(n_{13}) = \frac{r_1c_3}{n} = \frac{581(337)}{1786} = 109.63 \qquad \hat{E}(n_{23}) = \frac{r_2c_3}{n} = \frac{1205(337)}{1786} = 227.37$$

$$\hat{E}(n_{14}) = \frac{r_1c_4}{n} = \frac{581(372)}{1786} = 121.01 \qquad \hat{E}(n_{24}) = \frac{r_2c_4}{n} = \frac{1205(372)}{1786} = 250.99$$

The test statistic is $\chi^2 = \sum\sum \frac{\left[n_{ij} - \hat{E}(n_{ij})\right]^2}{\hat{E}(n_{ij})}$

$$= \frac{(117 - 169.49)^2}{169.49} + \frac{(222 - 180.87)^2}{180.87} + \frac{(133 - 109.63)^2}{109.63}$$
$$+ \frac{(109 - 121.01)^2}{121.01} + \frac{(404 - 351.51)^2}{351.51} + \frac{(334 - 375.13)^2}{375.13}$$
$$+ \frac{(204 - 227.37)^2}{227.37} + \frac{(263 - 250.99)^2}{250.99} = 47.11$$

The rejection region requires $\alpha = .05$ in the upper tail of the χ^2 distribution with df = $(r - 1)(c - 1) = (2 - 1)(4 - 1) = 3$. From Table XI, Appendix A, $\chi^2_{.05} = 7.81473$. The rejection region is $\chi^2 > 7.81473$.

Since the observed value of the test statistic falls in the rejection region ($\chi^2 = 47.11 > 7.81473$), H_0 is rejected. There is sufficient evidence to indicate a difference in the proportion of earthquake insured residents for the four sampled counties for $\alpha = .05$.

8.63 Some preliminary calculations are:

$$\hat{E}(n_{11}) = \frac{r_1c_1}{n} = \frac{202(432)}{1000} = 87.26 \qquad \hat{E}(n_{21}) = \frac{r_2c_1}{n} = \frac{247(432)}{1000} = 106.70$$

$$\hat{E}(n_{12}) = \frac{r_1c_2}{n} = \frac{202(457)}{1000} = 92.31 \qquad \hat{E}(n_{22}) = \frac{r_2c_2}{n} = \frac{247(457)}{1000} = 112.88$$

$$\hat{E}(n_{13}) = \frac{r_1c_3}{n} = \frac{202(111)}{1000} = 22.42 \qquad \hat{E}(n_{23}) = \frac{r_2c_3}{n} = \frac{247(111)}{1000} = 27.42$$

$$\hat{E}(n_{31}) = \frac{r_3c_1}{n} = \frac{234(432)}{1000} = 101.09 \qquad \hat{E}(n_{41}) = \frac{r_4c_1}{n} = \frac{198(432)}{1000} = 85.54$$

$$\hat{E}(n_{32}) = \frac{r_3c_2}{n} = \frac{234(457)}{1000} = 106.94 \qquad \hat{E}(n_{42}) = \frac{r_4c_2}{n} = \frac{198(457)}{1000} = 90.49$$

$$\hat{E}(n_{33}) = \frac{r_3c_3}{n} = \frac{234(111)}{1000} = 25.97 \qquad \hat{E}(n_{43}) = \frac{r_4c_3}{n} = \frac{198(111)}{1000} = 21.98$$

$$\hat{E}(n_{51}) = \frac{r_5c_1}{n} = \frac{119(432)}{1000} = 51.41$$

$$\hat{E}(n_{52}) = \frac{r_5c_2}{n} = \frac{119(457)}{1000} = 54.38$$

$$\hat{E}(n_{53}) = \frac{r_5c_3}{n} = \frac{119(111)}{1000} = 13.21$$

To determine if a relationship exists between age and number of games attended per year, we test:

H_0: Age and number of games attended per year are independent
H_a: Age and number of games attended per year are dependent

The test statistic is $\chi^2 = \sum\sum \frac{\left[n_{ij} - \hat{E}(n_{ij})\right]^2}{\hat{E}(n_{ij})} = \frac{(78 - 87.26)^2}{87.26} + \cdots + \frac{(22 - 13.21)^2}{13.21} = 103.08$

The rejection region requires $\alpha = .05$ in the upper tail of the χ^2 distribution with df = $(r - 1)(c - 1) = (5 - 1)(3 - 1) = 8$. From Table XI, Appendix A, $\chi^2_{.05} = 15.5073$. The rejection region is $\chi^2 > 15.5073$

Since the observed value of the test statistic falls in the rejection region ($\chi^2 = 103.08 > 15.5073$), H_0 is rejected. There is sufficient evidence to indicate a dependence between age and number of games attended per year at $\alpha = .05$.

8.65 Some preliminary calculations are:

If the proportions are all the same, then

$$p_{1,0} = p_{2,0} = p_{3,0} = p_{4,0} = .25$$

$$E(n_1) = E(n_2) = E(n_3) = E(n_4) = np_{i,0} = 151(.25) = 37.75$$

To determine if the proportions of accidents are different for at least two of the four time periods, we test:

H_0: $p_1 = p_2 = p_3 = p_4 = .25$
H_a: At least one $p_i \neq .25$

The test statistic is $\chi^2 = \sum \frac{[n_i - E(n_i)]^2}{E(n_i)}$

$$= \frac{(31 - 37.75)^2}{37.75} + \frac{(28 - 37.75)^2}{37.75} + \frac{(45 - 37.75)^2}{37.75} + \frac{(47 - 37.75)^2}{37.75} = 7.384$$

Since no α was given, we will use $\alpha = .05$. The rejection region requires $\alpha = .05$ in the upper tail of the χ^2 distribution with df $= k - 1 = 4 - 1 = 3$. From Table XI, Appendix A, $\chi^2_{.05} = 7.81473$. The rejection region is $\chi^2 > 7.81473$.

Since the observed value of the test statistic does not fall in the rejection region ($\chi^2 = 7.384 \not> 7.81473$), H_0 is not rejected. There is insufficient evidence to indicate the proportion of accidents are different for at least two of the four time periods at $\alpha = .05$.

8.67 To determine if there is a dependence between a son's choice of occupation and his father's occupation, we test:

H_0: Son's choice of occupation and his father's occupation are independent
H_a: Son's choice of occupation and his father's occupation are dependent.

The test statistic is $\chi^2 = \sum\sum \frac{[n_{ij} - \hat{E}(n_{ij})]^2}{\hat{E}(n_{ij})} = 180.874$ (from the printout).

The p-value is $p = 0.000$. Since the p-value is less than α ($p = 0.000 < .05$), H_0 is rejected. There is sufficient evidence to indicate a dependence between a son's choice of occupation and his father's occupation at $\alpha = .05$.

8.69 a. Some preliminary calculations are:

$$\hat{E}(n_{11}) = \frac{r_1c_1}{n} = \frac{30(30)}{200} = 4.5 \qquad \hat{E}(n_{12}) = \frac{r_1c_2}{n} = \frac{30(120)}{200} = 18$$

$$\hat{E}(n_{13}) = \frac{r_1c_3}{n} = \frac{30(50)}{200} = 7.5 \qquad \hat{E}(n_{21}) = \frac{r_2c_1}{n} = \frac{75(30)}{200} = 11.25$$

$$\hat{E}(n_{22}) = \frac{r_2c_2}{n} = \frac{75(120)}{200} = 45 \qquad \hat{E}(n_{23}) = \frac{r_2c_3}{n} = \frac{75(50)}{200} = 18.75$$

$$\hat{E}(n_{31}) = \frac{r_3c_1}{n} = \frac{65(30)}{200} = 9.75 \qquad \hat{E}(n_{32}) = \frac{r_3c_2}{n} = \frac{65(120)}{200} = 39$$

$$\hat{E}(n_{33}) = \frac{r_3c_3}{n} = \frac{65(50)}{200} = 16.25 \qquad \hat{E}(n_{41}) = \frac{r_4c_1}{n} = \frac{30(30)}{200} = 4.5$$

$$\hat{E}(n_{42}) = \frac{r_4c_2}{n} = \frac{30(120)}{200} = 18 \qquad \hat{E}(n_{43}) = \frac{r_4c_3}{n} = \frac{30(50)}{200} = 7.5$$

To determine if a relationship exists between family status and the use of alcohol, we test:

H_0: Family status and use of alcohol are independent
H_a: Family status and use of alcohol are dependent

The test statistic is $\chi^2 = \sum\sum \frac{[n_{ij} - \hat{E}(n_{ij})]^2}{\hat{E}(n_{ij})}$

$$= \frac{(4 - 4.5)^2}{4.5} + \frac{(11 - 11.25)^2}{11.25} + \frac{(9 - 9.75)^2}{9.75} + \cdots + \frac{(7 - 18)^2}{18} = 8.664$$

The rejection region requires $\alpha = .05$ in the upper tail of the χ^2 distribution with df $= (r - 1)(c - 1) = (4 - 1)(3 - 1) = 6$. From Table XI, Appendix A, $\chi^2_{.05} = 12.5916$. The rejection region is $\chi^2 > 12.5916$.

Since the observed value of the test statistic does not fall in the rejection region ($\chi^2 = 8.664 \not> 12.5916$), H_0 is not rejected. There is insufficient evidence to indicate a relationship exists between family status and the use of alcohol at $\alpha = .05$.

b. We must assume:

1. The n observed counts are a random sample from the population of interest. We may consider this a multinomial experiment with $4 \times 3 = 12$ possible outcomes.
2. The sample size, n, will be large enough so that, for every cell count, $E(n_{ij})$ will be equal to 5 or more.

The last assumption is violated since we had 2 cells with expected counts of less than 5.

c. Suppose family status is the base variable. To find the cell percentages, divide the number in each cell by its corresponding row total and multiply by 100%. To find the column percentages, divide the column totals by the total sample size and multiply by 100%.

The percentage in the first cell of row one is $4/30 \times 100\% = 13.3\%$. The other percentages are found in a similar manner and are shown in the table.

		Alcohol			
		None	Occasional	Frequent	Total
Family Status	Upper Class	13.3%	53.3%	33.3%	100
	Upper Middle Class	14.7%	53.3%	32.0%	100
	Lower Middle Class	13.8%	72.3%	13.8%	100
	Lower Class	20.0%	56.7%	23.3%	100
		15.0%	60.0%	25.0%	100

The graph is:

Yes, the graph supports the results in part a. The graphs for each level of alcohol do not cross each other.

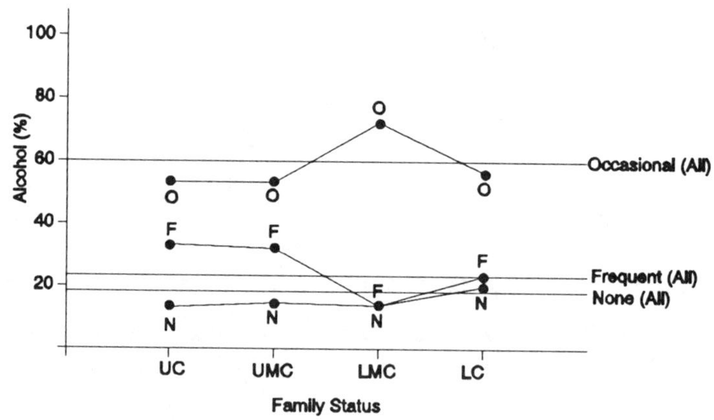

8.71 Some preliminary calculations are:

$$\hat{E}(n_{11}) = \frac{r_1c_1}{n} = \frac{52(45)}{100} = 23.4 \qquad \hat{E}(n_{21}) = \frac{r_2c_1}{n} = \frac{48(45)}{100} = 21.6$$

$$\hat{E}(n_{12}) = \frac{r_1c_2}{n} = \frac{52(55)}{100} = 28.6 \qquad \hat{E}(n_{22}) = \frac{r_2c_2}{n} = \frac{48(55)}{100} = 26.4$$

To determine if a relationship exists between eye and hair color, we test:

H_0: Hair and eye color are independent
H_a: Hair and eye color are dependent

The test statistic is $\chi^2 = \sum\sum \frac{\left[n_{ij} - \hat{E}(n_{ij})\right]^2}{\hat{E}(n_{ij})}$

$$= \frac{(31 - 23.4)^2}{23.4} + \frac{(21 - 28.6)^2}{28.6} + \frac{(14 - 21.6)^2}{21.6} + \frac{(34 - 26.4)^2}{26.4} = 9.35$$

The rejection region requires $\alpha = .10$ in the upper tail of the χ^2 distribution with df = $(r - 1)(c - 1) = (2 - 1)(2 - 1) = 1$. From Table XI, Appendix A, $\chi^2_{.10} = 2.70554$. The rejection region is $\chi^2 > 2.70554$.

Since the observed value of the test statistic falls in the rejection region ($\chi^2 = 9.35 > 2.70554$), H_0 is rejected. There is sufficient evidence to indicate that hair and eye color are related at $\alpha = .10$.

8.73 Some preliminary calculations are:

$E(n_1) = np_1 = 144(.25) = 36$
$E(n_2) = np_2 = 144(.5) = 72$
$E(n_3) = np_3 = 144(.25) = 36$

To determine if there is evidence to contradict the geneticist's theory, we test:

H_0: $p_1 = .25$, $p_2 = .50$, and $p_3 = .25$
H_a: At least one of the probabilities differs from its hypothesized value

The test statistic is $\chi^2 = \sum \frac{\left[n_i - E(n_i)\right]^2}{E(n_i)} = \frac{(30 - 36)^2}{36} + \frac{(78 - 72)^2}{72} + \frac{(36 - 36)^2}{36} = 1.5$

Since no α was given, we will use $\alpha = .05$. The rejection region requires $\alpha = .05$ in the upper tail of the χ^2 distribution with df $= k - 1 = 3 - 1 = 2$. From Table XI, Appendix A, $\chi^2_{.05} = 5.99147$. The rejection region is $\chi^2 > 5.99147$.

Since the observed value of the test statistic does not fall in the rejection region ($\chi^2 = 1.5 \ngtr 5.99147$), H_0 is not rejected. There is insufficient evidence to contradict the geneticist's theory at $\alpha = .05$.

Simple Linear Regression

Chapter 9

9.1 a.

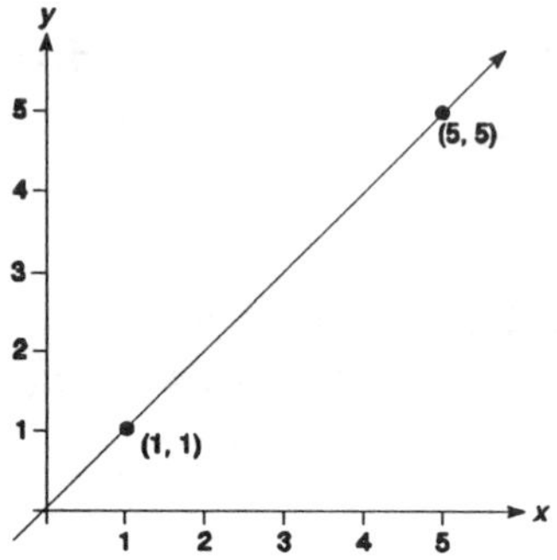

b.

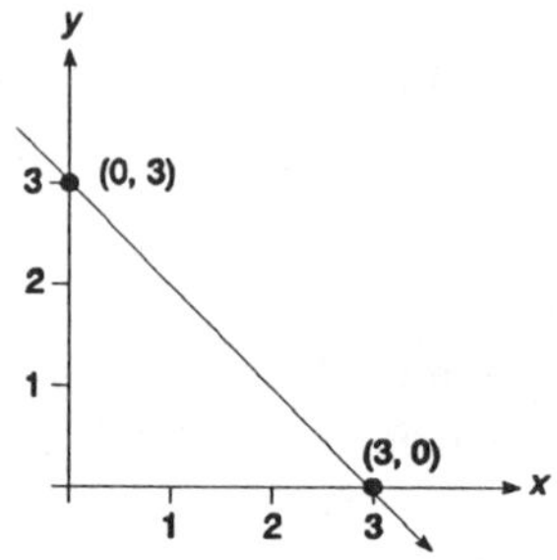

c.

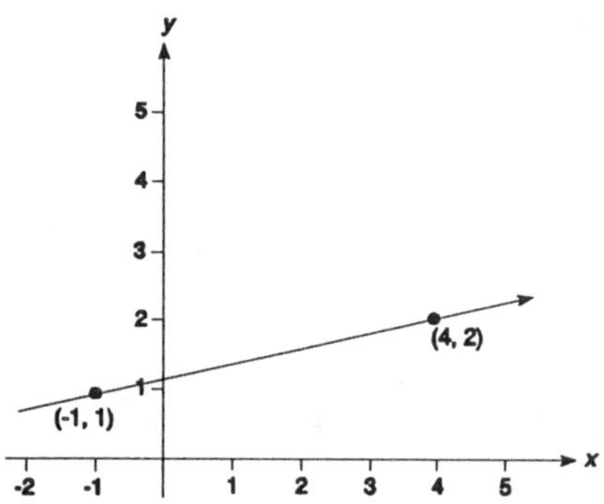

d.

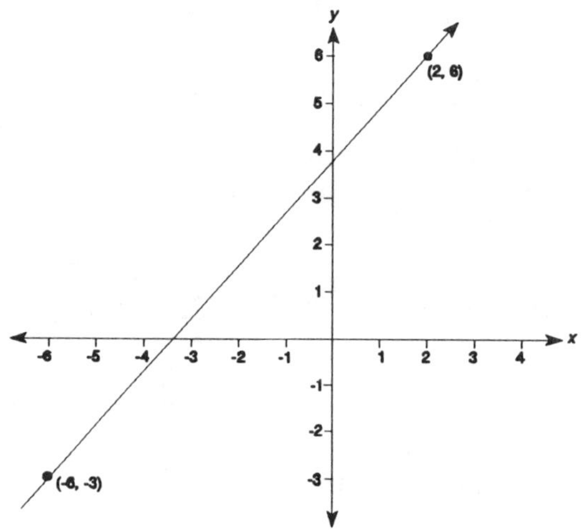

9.3 The two equations are:

$$4 = \beta_0 + \beta_1(-2) \text{ and } 6 = \beta_0 + \beta_1(4)$$

Subtracting the first equation from the second, we get:

$$\begin{array}{rl} 6 = & \beta_0 + 4\beta_1 \\ -(4 = & \beta_0 - 2\beta_1) \\ \hline 2 = & \quad 6\beta_1 \end{array} \Rightarrow \beta_1 = \frac{2}{6} = \frac{1}{3}$$

Substituting $\beta_1 = \frac{1}{3}$ into the first equation, we get:

$$4 = \beta_0 + \frac{1}{3}(-2) \Rightarrow \beta_0 = 4 + \frac{2}{3} = \frac{14}{3}$$

The equation for the line is $y = \frac{14}{3} + \frac{1}{3}x$.

9.5 To graph a line, we need two points. Pick two values for x, and find the corresponding y values by substituting the values of x into the equation.

a. Let $x = 0 \Rightarrow y = 4 + (0) = 4$
and $x = 2 \Rightarrow y = 4 + (2) = 6$

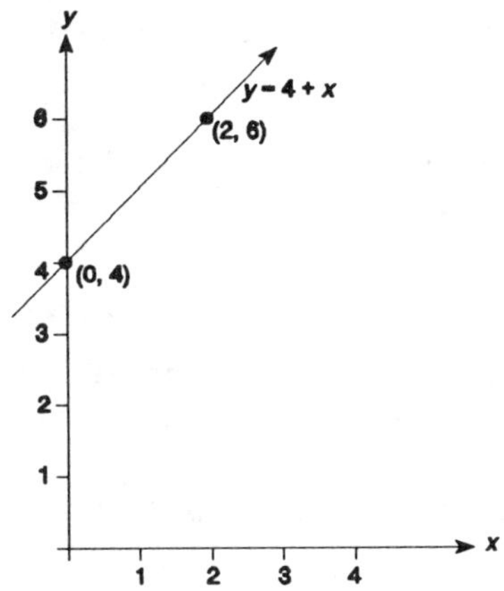

b. Let $x = 0 \Rightarrow y = 5 - 2(0) = 5$
and $x = 2 \Rightarrow y = 5 - 2(2) = 1$

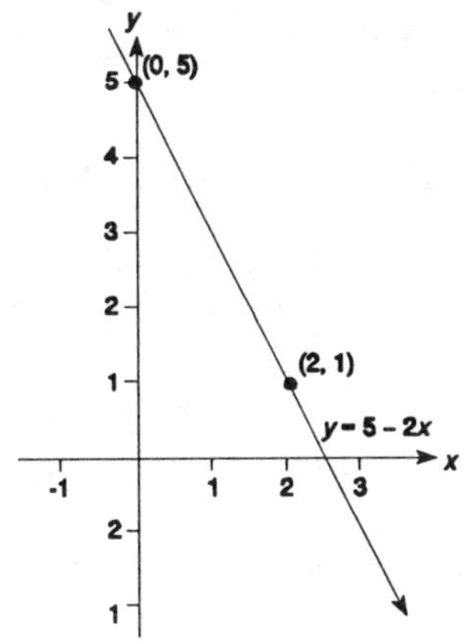

c. Let $x = 0 \Rightarrow y = -4 + 3(0) = -4$
and $x = 2 \Rightarrow y = -4 + 3(2) = 2$

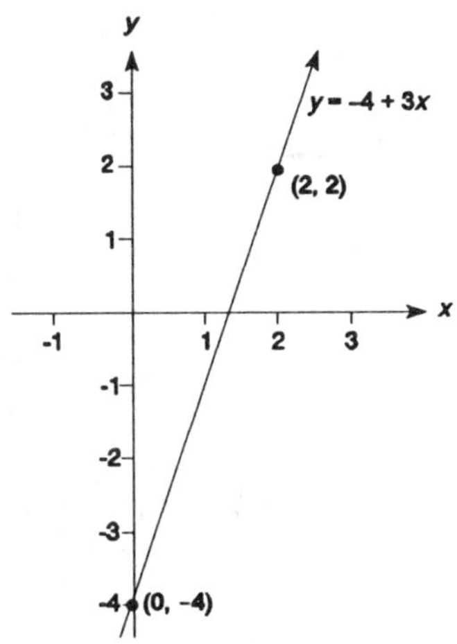

d. Let $x = 0 \Rightarrow y = -2(0) = 0$
and $x = 2 \Rightarrow y = -2(2) = -4$

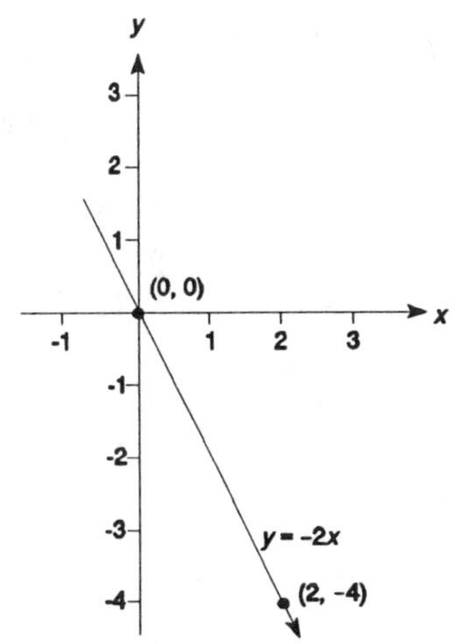

e. Let $x = 0 \Rightarrow y = 0$
and $x = 2 \Rightarrow y = 2$

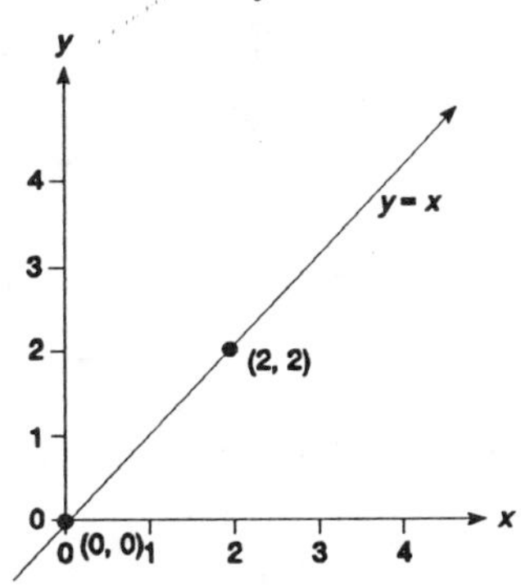

f. Let $x = 0 \Rightarrow y = .5 + 1.5(0) = .5$
and $x = 2 \Rightarrow y = .5 + 1.5(2) = 3.5$

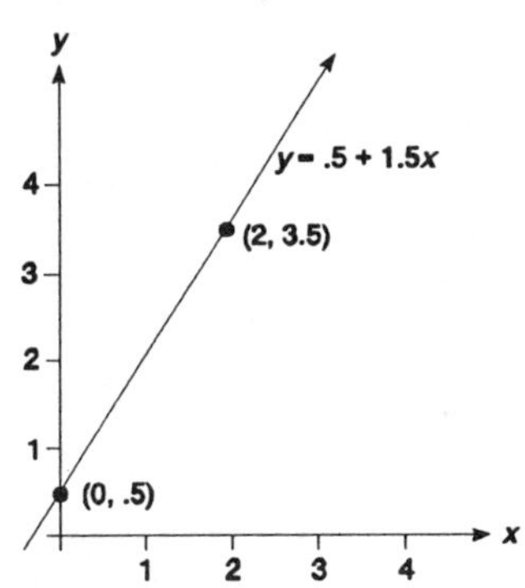

9.7 A deterministic model does not allow for random error or variation, whereas a probabilistic model does. An example where a deterministic model would be appropriate is:

Let y = cost of a 2 × 4 piece of lumber and
x = length (in feet)

The model would be $y = \beta_1 x$. There should be no variation in price for the same length of wood.

An example where a probabilistic model would be appropriate is:

Let y = sales per month of a commodity and
x = amount of money spent advertising

The model would be $y = \beta_0 + \beta_1 x + \epsilon$. The sales per month will probably vary even if the amount of money spent on advertising remains the same.

9.9 No. The random error component, ϵ, allows the values of the variable to fall above or below the line.

9.11 From Exercise 9.10, $\hat{\beta}_0 = 7.10$ and $\hat{\beta}_1 = -.78$.

The fitted line is $\hat{y} = 7.10 - .78x$. To obtain values for $\hat{y}$, we substitute values of x into the equation and solve for $\hat{y}$.

a.

x	y	$\hat{y} = 7.10 - .78x$	$(y - \hat{y})$	$(y - \hat{y})^2$
7	2	1.64	.36	.1296
4	4	3.98	.02	.0004
6	2	2.42	−.42	.1764
2	5	5.54	−.54	.2916
1	7	6.32	.68	.4624
1	6	6.32	−.32	.1024
3	5	4.76	.24	.0576
			$\sum(y - \hat{y}) = 0.02$	SSE = $\sum(y - \hat{y})^2 = 1.2204$

b.

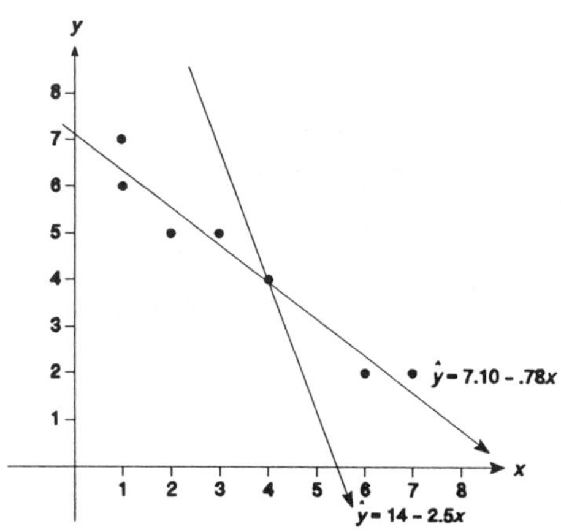

c.

x	y	$\hat{y} = 14 - 2.5x$	$(y - \hat{y})$	$(y - \hat{y})^2$
7	2	−3.5	5.5	30.25
4	4	4	0	0
6	2	−1	3	9
2	5	9	−4	16
1	7	11.5	−4.5	20.25
1	6	11.5	−5.5	30.25
3	5	6.5	−1.5	2.25
			$\sum(y - \hat{y}) = -7$	SSE = 108.00

9.13 a.

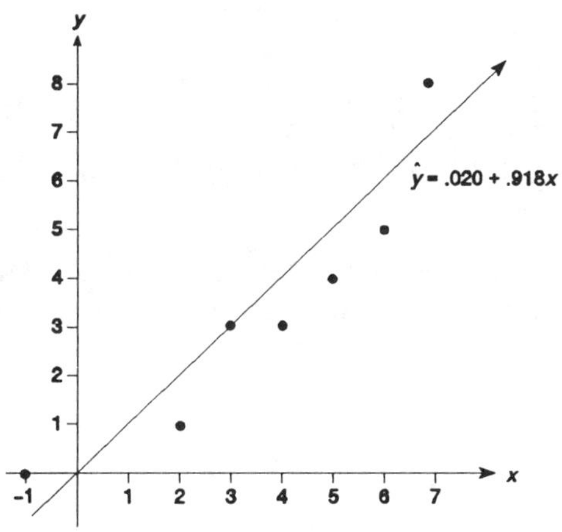

b. As x increases, y tends to increase. Thus, there appears to be a positive, linear relationship between y and x.

c. $\hat{\beta}_1 = \dfrac{SS_{xy}}{SS_{xx}} = \dfrac{39.8571}{43.4286} = 9.177616 \approx .918$

$\hat{\beta}_0 = \bar{y} - \hat{\beta}_1 \bar{x} = 3.4286 - .9177616(3.7143) = .0197581 \approx .020$

d. The line appears to fit the data quite well.

e. $\hat{\beta}_0 = .020$ The estimated mean value of y when $x = 0$ is .020.
$\hat{\beta}_1 = .918$ The estimated change in the mean value of y for each unit change in x is .918.

These interpretations are valid only for values of x in the range from -1 to 7.

9.15 a. Yes. The response variable is verbal memory retention. Small values of verbal memory retention correspond with deficits in short-term verbal memory. From the plot, it appears that as the hippocampal volume gets smaller, the verbal memory retention also gets smaller.

b. A straight line model would be $y = \beta_0 + \beta_1 x + \epsilon$.

c. We would expect the slope of the line to be positive. As hippocampal volume increases, the verbal memory retention should also get larger.

9.17 a. A scattergram of the data is:

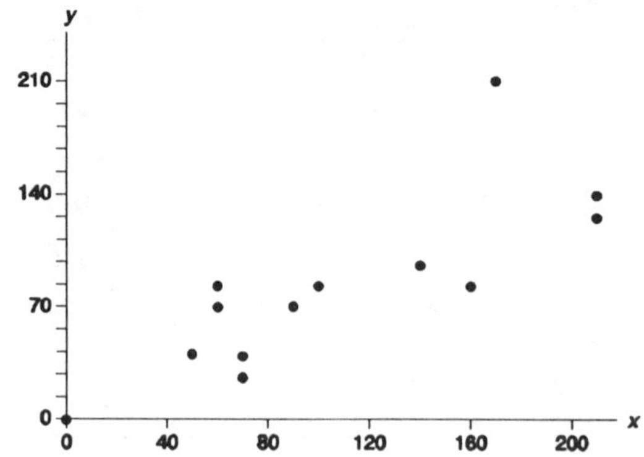

It appears that there is a positive linear relationship between total number of words with unfamiliar partner and with familiar partner.

b. A straight line model would be $y = \beta_0 + \beta_1 x + \epsilon$.

c. $\hat{\beta}_0 = 20.127511$ and $\hat{\beta}_1 = .624416$.

d. $\hat{\beta}_0 = 20.127511$. Since $x = 0$ is in the observed range, the mean number of words in conversation with an unfamiliar partner is estimated to be 20.127511 when the number of words in conversation with a familiar partner is 0.

$\hat{\beta}_1 = .624416$. For each additional word in conversation with a familiar partner, the mean number of words in conversation with an unfamiliar partner is estimated to increase by .624416.

9.19 a. Some preliminary calculations are:

$$\sum x_i = 438 \qquad \sum y_i = 50 \qquad n = 7$$

$$\sum x_i^2 = 27{,}582 \qquad \sum x_i y_i = 2{,}986$$

$$SS_{xy} = \sum x_i y_i - \frac{\left(\sum x_i\right)\left(\sum y_i\right)}{n} = 2{,}986 - \frac{438(50)}{7} = -142.571429$$

$$SS_{xx} = \sum x_i^2 - \frac{\left(\sum x_i\right)^2}{n} = 27{,}582 - \frac{438^2}{7} = 175.71429$$

$$\hat{\beta}_1 = \frac{SS_{xy}}{SS_{xx}} = \frac{-142.571429}{175.71429} = -.811382096 \approx -.8114$$

$$\hat{\beta}_0 = \bar{y} - \hat{\beta}_1 \bar{x} = \frac{50}{7} - (-.811382096)\frac{438}{7} = 57.91219404 \approx 57.9122$$

The least squares line is $\hat{y} = 57.9122 - .8114x$.

b. The scattergram of the data is:

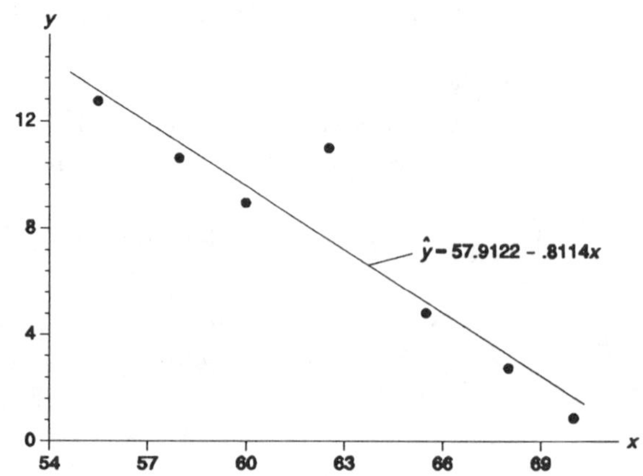

c. For $x = 60$, $\hat{y} = 57.9122 - .8114(60) = 9.2282$.

9.21 a. $\hat{\beta}_1 = \frac{SS_{xy}}{SS_{xx}} = \frac{100.0825}{83.3425} = 1.200857906 \approx 1.201$

$\hat{\beta}_0 = \bar{y} - \hat{\beta}_1\bar{x} = 8.5083 - 1.200857906(4.2750) = 3.374632452 \approx 3.375$

The least squares equation is $\hat{y} = 3.375 + 1.201x$.

b. The line appears to fit the data fairly well.

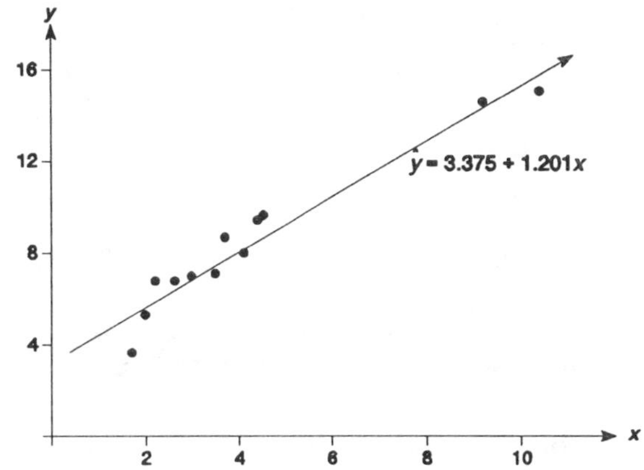

c. $\hat{\beta}_0 = 3.375$. Since outside air temperature $= 0$ is not in the observed range, $\hat{\beta}_0$ has no meaning.

$\hat{\beta}_1 = 1.201$. For each 1° increase in outside air temperature, the mean cocoon temperature will increase by an estimated 1.201°.

9.23 a. $SSE = SS_{yy} - \hat{\beta}_1 SS_{xy} = 95 - .75(50) = 57.5$

$s^2 = \frac{SSE}{n-2} = \frac{57.5}{20-2} = 3.19444$

b. $SS_{yy} = \sum y^2 - \frac{(\sum y)^2}{n} = 860 - \frac{50^2}{40} = 797.5$

$SSE = SS_{yy} - \hat{\beta}_1 SS_{xy} = 797.5 - .2(2700) = 257.5$

$s^2 = \frac{SSE}{n-2} = \frac{257.5}{40-2} = 6.776315789 \approx 6.7763$

c. $SS_{yy} = \sum (y_i - \bar{y})^2 = 58 \qquad \hat{\beta}_1 = \frac{SS_{xy}}{SS_{xx}} = \frac{91}{170} = .535294117$

$SSE = SS_{yy} - \hat{\beta}_1 SS_{xy} = 58 - .535294117(91) = 9.2882353 \approx 9.288$

$s^2 = \frac{SSE}{n-2} = \frac{9.2882353}{10-2} = 1.161029413 \approx 1.1610$

9.25 The graph in **b** would have the smallest s^2 because the width of the data points is the smallest.

9.27 a. $SSE = 2.7091$, $s^2 = MSE = 0.1693$, and $s = 0.4115$.

b. We would expect most of the observed best jumping distance values to fall within $2s$ or $2(.4115)$ or .823 meters of their predicted values.

9.29 a. SSE = 14.31188, s^2 = MSE = 1.19266, and s = Root MSE = 1.09209

b. We would expect most of the observed number of flycatchers killed to fall within $2s$ or 2(1.09209) or 2.184 flycatchers of their predicted values.

9.31 a.

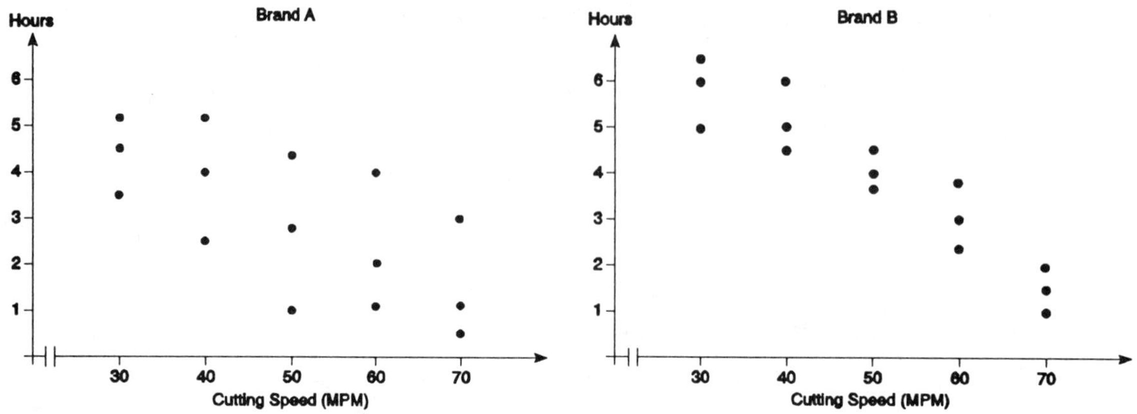

b. For Brand A,

$$\sum x_i = 750 \quad \sum y_i = 44.8 \quad \sum x_i y_i = 2022 \quad \sum x_i^2 = 40{,}500 \quad \sum y_i^2 = 168.7$$

$$SS_{xx} = \sum x_i^2 - \frac{\left(\sum x_i\right)^2}{n} = 40{,}500 - \frac{750^2}{15} = 40{,}500 - 37{,}500 = 3000$$

$$SS_{xy} = \sum x_i y_i - \frac{\left(\sum x_i\right)\left(\sum y_i\right)}{n} = 2022 - \frac{(750)(44.8)}{15} = 2022 - 2240 = -218$$

$$\hat{\beta}_1 = \frac{SS_{xy}}{SS_{xx}} = \frac{-218}{3000} = -.07266667 \approx -.0727$$

$$\hat{\beta}_0 = \bar{y} - \hat{\beta}_1 \bar{x} = \frac{44.8}{15} - (-.07266667)\left(\frac{750}{15}\right) = 2.9866667 + 3.633333 = 6.62$$

$$\hat{y} = 6.62 - .0727x$$

For Brand B,

$$\sum x_i = 750 \quad \sum y_i = 58.9 \quad \sum x_i y_i = 2622 \quad \sum x_i^2 = 40{,}500 \quad \sum y_i^2 = 270.89$$

$$SS_{xx} = \sum x_i^2 - \frac{\left(\sum x_i\right)^2}{n} = 40{,}500 - \frac{(750)^2}{15} = 40{,}500 - 37{,}500 = 3000$$

$$SS_{xy} = \sum xy - \frac{\left(\sum x\right)\left(\sum y\right)}{n} = 2622 - \frac{(750)(58.9)}{15} = 2622 - 2945 = -323$$

$$\hat{\beta}_1 = \frac{SS_{xy}}{SS_{xx}} = \frac{-323}{3000} = -.10766667 \approx -.1077$$

$$\hat{\beta}_0 = \bar{y} - \hat{\beta}_1 \bar{x} = \left(\frac{59.9}{15}\right) - (-.10766667)\left(\frac{750}{15}\right) = 3.92667 + 5.38333 = 9.31$$

$$\hat{y} = 9.31 - .1077x$$

c. For Brand A,

$$SS_{yy} = \sum y_i^2 - \frac{\left(\sum y_i\right)^2}{n} = 168.7 - \frac{(44.8)^2}{15} = 168.7 - 133.802667 = 34.8973333$$

$$SSE = SS_{yy} - \hat{\beta}_1 SS_{xy} = 34.8973333 - (-.07266667)(-218)$$
$$= 34.8973333 - 15.8413333 = 19.056$$

$$s^2 = \frac{SSE}{n-2} = \frac{19.056}{13} = 1.465846154 \qquad s = \sqrt{1.465846154} = 1.211$$

For Brand B,

$$SS_{yy} = \sum y_i^2 - \frac{\left(\sum y_i\right)^2}{n} = 270.89 - \frac{(58.9)^2}{15} = 270.89 - 231.2806667$$
$$= 39.6093333$$

$$SSE = SS_{yy} - \hat{\beta}_1 SS_{xy} = 39.6093333 - (-.10766667)(-323)$$
$$= 39.6093333 - 34.7763333 = 4.83299999 \approx 4.833$$

$$s^2 = \frac{SSE}{n-2} = \frac{4.833}{13} = .37176923 \qquad s = \sqrt{.37176923} = .610$$

d. For Brand A,

$\hat{y} = 6.62 - .0727x$. For $x = 70$, $\hat{y} = 6.62 - .0727(70) = 1.531$

$2s = 2(1.211) = 2.422$

Therefore, $\hat{y} \pm 2s \Rightarrow 1.531 \pm 2.422 \Rightarrow (-.891, 3.593)$

For Brand B, $\hat{y} = 9.31 - .1077x$. For $x = 70$, $\hat{y} = 9.31 - .1077(70) = 1.771$

$2s = 2(.61) = 1.22$

Therefore, $\hat{y} \pm 2s \Rightarrow 1.771 \pm 1.22 \Rightarrow (.551, 2.991)$

e. More confident with Brand B since there is less variation (s is smaller).

9.33 a.

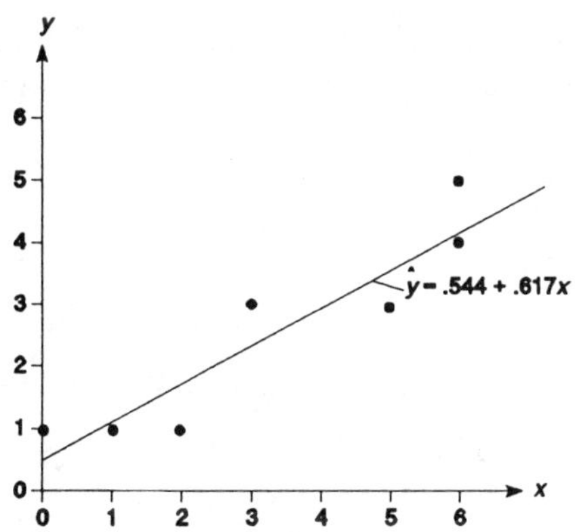

b. Some preliminary calculations are:

$\sum x = 23 \qquad \sum x^2 = 111 \qquad \sum xy = 81$

$\sum y = 18 \qquad \sum y^2 = 62$

$$SS_{xy} = \sum xy - \frac{\sum x \sum y}{n} = 81 - \frac{23(18)}{7} = 21.85714286$$

$$SS_{xx} = \sum x^2 - \frac{(\sum x)^2}{n} = 111 - \frac{23^2}{7} = 35.42857143$$

$$SS_{yy} = \sum y^2 - \frac{(\sum y)^2}{n} = 62 - \frac{18^2}{7} = 15.71428571$$

$$\hat{\beta}_1 = \frac{SS_{xy}}{SS_{xx}} = \frac{21.85714286}{35.42857143} = .616935483 \approx .617$$

$$\hat{\beta}_0 = \bar{y} - \hat{\beta}_1 \bar{x} = \frac{18}{7} - .616935483\frac{23}{7} = .544354838 \approx .544$$

The least squares line is $\hat{y} = .544 + .617x$.

c. The line is plotted on the graph in a.

d. To determine if x contributes information for the linear prediction of y, we test:

H_0: $\beta_1 = 0$
H_a: $\beta_1 \neq 0$

e. The test statistic is $t = \dfrac{\hat{\beta}_1 - 0}{\dfrac{s}{\sqrt{SS_{xx}}}} = \dfrac{.617 - 0}{\dfrac{.6678}{\sqrt{35.42857143}}} = 5.50$

where $SSE = SS_{yy} - \hat{\beta}_1 SS_{xy} = 15.71428571 - .616935483(21.85714286) = 2.22983872$

$s^2 = \dfrac{SSE}{n - 2} = \dfrac{2.22983872}{7 - 2} = .44596774 \qquad s = \sqrt{.44596774} = .6678$

The degrees of freedom are $n - 2 = 7 - 2 = 5$.

f. The rejection region requires $\alpha/2 = .05/2 = .025$ in each tail of the t-distribution with df = 5. From Table IV, Appendix A, $t_{.025} = 2.571$. The rejection region is $t < -2.571$ or $t > 2.571$.

Since the observed value of the test statistic falls in the rejection region ($t = 5.50 > 2.571$), H_0 is rejected. There is sufficient evidence to indicate x contributes information for the linear prediction of y at $\alpha = .05$.

9.35 Some preliminary calculations are:

$\sum x = 19 \qquad \sum x^2 = 71 \qquad \sum xy = 66$
$\sum y = 20 \qquad \sum y^2 = 74$

$$SS_{xy} = \sum xy - \frac{\sum x \sum y}{n} = 66 - \frac{19(20)}{6} = 2.66666667$$

$$SS_{xx} = \sum x^2 - \frac{(\sum x)^2}{n} = 71 - \frac{19^2}{6} = 10.83333333$$

$$SS_{yy} = \sum y^2 - \frac{(\sum y)^2}{n} = 74 - \frac{20^2}{6} = 7.33333333$$

$$\hat{\beta}_1 = \frac{SS_{xy}}{SS_{xx}} = \frac{2.66666667}{10.83333333} = .246153846 \approx .246$$

$SSE = SS_{yy} - \hat{\beta}_1 SS_{xy} = 7.33333333 - (.246153846)(2.66666667) = 6.676923073$

$$s^2 = \frac{SSE}{n-2} = \frac{6.676923073}{6-2} = 1.669230768 \qquad s = \sqrt{1.669230768} = 1.2920$$

To determine if a straight line is useful for characterizing the relationship between x and y, we test:

H_0: $\beta_1 = 0$
H_a: $\beta_1 \neq 0$

The test statistic is $t = \dfrac{\hat{\beta}_1 - 0}{s_{\hat{\beta}_1}} = \dfrac{.246}{\dfrac{1.292}{\sqrt{10.83333333}}} = .627$

The rejection region requires $\alpha/2 = .05/2 = .025$ in each tail of the t-distribution with df $= n - 2 = 6 - 2 = 4$. From Table IV, Appendix A, $t_{.025} = 2.776$. The rejection region is $t < -2.776$ or $t > 2.776$.

Since the observed value of the test statistic does not fall in the rejection region ($t = .627 \not> 2.776$), H_0 is not rejected. There is insufficient evidence to indicate a straight line is useful for characterizing the relationship between x and y at $\alpha = .05$.

9.37 a. The scattergram of the data is:

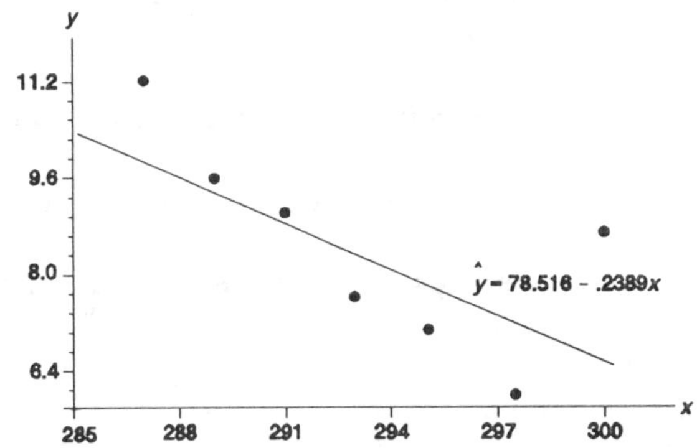

b. Some preliminary calculations are:

$\sum x_i = 2{,}053 \qquad \sum y_i = 59.2 \qquad n = 7$

$\sum x_i^2 = 602{,}265.5 \qquad \sum y_i^2 = 517.22 \qquad \sum x_i y_i = 17{,}326.7$

$$SS_{xy} = \sum x_i y_i - \frac{\left(\sum x_i\right)\left(\sum y_i\right)}{n} = 17{,}326.7 - \frac{2053(59.2)}{7} = -35.81429$$

$$SS_{xx} = \sum x_i^2 - \frac{\left(\sum x_i\right)^2}{n} = 602{,}265.5 - \frac{2053^2}{7} = 149.9286$$

$$SS_{yy} = \sum y_i^2 - \frac{\left(\sum y_i\right)^2}{n} = 517.22 - \frac{59.2^2}{7} = 16.5571429$$

$$\hat{\beta}_1 = \frac{SS_{xy}}{SS_{xx}} = \frac{-35.81429}{149.9286} = -.238875638 \approx -.2389$$

$$\hat{\beta}_0 = \bar{y} - \hat{\beta}_1\bar{x} = \frac{59.2}{7} - (-.238875638)\frac{2053}{7} = 78.515955 \approx 78.516$$

The least squares line is $\hat{y} = 78.516 - .2389x$.

c. The least squares line is plotted on the graph in part a.

d. Some additional calculations:

$$SSE = SS_{yy} - \hat{\beta}_1 SS_{xy} = 16.5571429 - (-.238875638)(-35.81429) = 8.001980951$$

$$s^2 = \frac{SSE}{n-2} = \frac{8.001980951}{7-2} = 1.60039619 \qquad s = \sqrt{1.60039619} = 1.2651$$

To determine if the model is adequate, we test:

H_0: $\beta_1 = 0$
H_a: $\beta_1 \neq 0$

The test statistic is $t = \dfrac{\hat{\beta}_1 - 0}{s_{\hat{\beta}_1}} = \dfrac{-.2389 - 0}{\dfrac{1.2651}{\sqrt{149.9286}}} = -2.31$

The rejection region requires $\alpha/2 = .01/2 = .005$ in each tail of the t-distribution with df $= n - 2 = 7 - 2 = 5$. From Table IV, Appendix A, $t_{.005} = 4.032$. The rejection region is $t < -4.032$ or $t > 4.032$.

Since the observed value of the test statistic does not fall in the rejection region ($t = -2.31 \nless -4.032$), H_0 is not rejected. There is insufficient evidence to indicate that the model is adequate at $\alpha = .01$.

e. One point looks like a possible outlier: (301, 8.5).

f. We will redo the analysis after eliminating the above point:

(a) The scattergram of the data is:

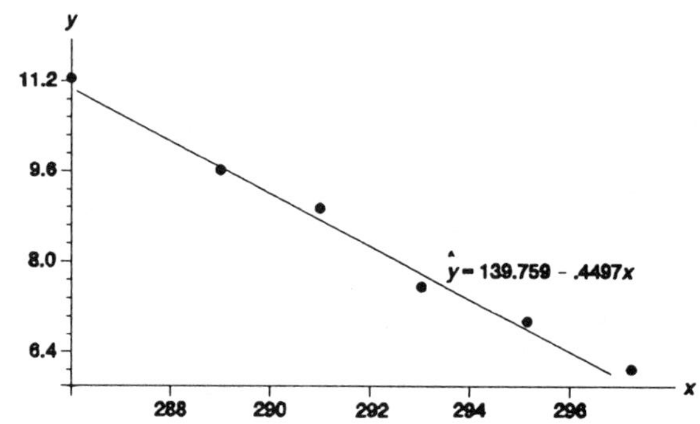

(b) Some preliminary calculations are:

$$\sum x_i = 1{,}752 \qquad \sum y_i = 50.7 \qquad n = 6$$

$$\sum x_i^2 = 511{,}664.5 \qquad \sum y_i^2 = 444.97 \qquad \sum x_i y_i = 14{,}768.2$$

$$SS_{xy} = \sum x_i y_i - \frac{\left(\sum x_i\right)\left(\sum y_i\right)}{n} = 14{,}768.2 - \frac{1752(50.7)}{6} = -36.2$$

$$SS_{xx} = \sum x_i^2 - \frac{\left(\sum x_i\right)^2}{n} = 511{,}664.5 - \frac{1752^2}{6} = 80.5$$

$$SS_{yy} = \sum y_i^2 - \frac{\left(\sum y_i\right)^2}{n} = 444.97 - \frac{50.7^2}{6} = 16.555$$

$$\hat{\beta}_1 = \frac{SS_{xy}}{SS_{xx}} = \frac{-36.2}{80.5} = -.449689441 \approx -.4497$$

$$\hat{\beta}_0 = \bar{y} - \hat{\beta}_1\bar{x} = \frac{50.7}{6} - (-.449689441)\frac{1752}{6} = 139.7593168 \approx 139.759$$

The least squares line is $\hat{y} = 139.759 - .4497x$.

(c) The least squares line is plotted on the graph in part **f(a)**.

(d) Some additional calculations:

$$SSE = SS_{yy} - \hat{\beta}_1 SS_{xy} = 16.555 - (-.449689441)(-36.2) = .27624224$$

$$s^2 = \frac{SSE}{n-2} = \frac{.27624224}{6-2} = 0.06906056 \qquad s = \sqrt{0.06906056} = 0.2628$$

To determine if the model is adequate, we test:

H_0: $\beta_1 = 0$
H_a: $\beta_1 \neq 0$

The test statistic is $t = \dfrac{\hat{\beta}_1 - 0}{s_{\hat{\beta}_1}} = \dfrac{-.4497 - 0}{\frac{0.2628}{\sqrt{80.5}}} = -15.35$

The rejection region requires $\alpha/2 = .01/2 = .005$ in each tail of the t-distribution with df $= n - 2 = 6 - 2 = 4$. From Table IV, Appendix A, $t_{.005} = 4.604$. The rejection region is $t < -4.604$ or $t > 4.604$.

Since the observed value of the test statistic falls in the rejection region ($t = -15.35 < -4.604$), H_0 is rejected. There is sufficient evidence to indicate that the model is adequate at $\alpha = .01$.

9.39 a. To determine if x and y are linearly related, we test:

H_0: $\beta_1 = 0$
H_a: $\beta_1 \neq 0$

The test statistic is $t = 4.98$.

The p-value is .001. Since the p-value is less than $\alpha = .01$, H_0 is rejected at $\alpha = .01$. There is sufficient evidence to indicate that x and y are linearly related.

b. Since the model is adequate, it is reasonable to use it to predict values of y.

For $x = 4$, $\hat{y} = .202 + .135x = .202 + .135(3) = .607$. This value is meaningful only if $x = 3$ is within the observed range.

9.41 a. To determine if y is positively linearly related to x, we test:

H_0: $\beta_1 = 0$
H_a: $\beta_1 > 0$

The test statistic is $t = \dfrac{\hat{\beta}_1 - 0}{s_{\hat{\beta}_1}} = \dfrac{.1077 - 0}{\dfrac{0.2629}{\sqrt{149.9286}}} = 4.004$

The rejection region requires $\alpha = .01$ in the upper tail of the t-distribution with df $= n - 2 = 14 - 2 = 12$. From Table IV, Appendix A, $t_{.01} = 2.681$. The rejection region is $t > 2.681$.

Since the observed value of the test statistic falls in the rejection region ($t = 4.004 > 2.681$), H_0 is rejected. There is sufficient evidence to indicate that y is positively linearly related to x at $\alpha = .01$.

b. For confidence level .99, $\alpha = .01$ and $\alpha/2 = .01/2 = .005$. From Table IV, Appendix A, with df $= n - 2 = 14 - 2 = 12$, $t_{.005} = 3.055$. The confidence interval is:

$$\hat{\beta}_1 \pm t_{.005} s_{\hat{\beta}_1} \Rightarrow .1077 \pm 3.055(.0269) \Rightarrow .1077 \pm .0822 \Rightarrow (.0255, .1899)$$

We are 99% confident that the increase in the mean number of flycatchers that are killed for each additional nest box nit increase is between .0255 and .1889. This implies that as the nest box nit occupancy increases, the number of flycatchers killed also increases.

9.43 a. From Exercise 9.21, $\hat{\beta}_1 = 1.200857906 \approx 1.201$

$$SSE = SS_{yy} - \hat{\beta}_1 SS_{xy} = 127.5092 - (1.200857906)(100.0825) = 7.3243386$$

$$s^2 = \frac{SSE}{n - 2} = \frac{7.3243386}{12 - 2} = .73243386 \qquad s = \sqrt{.73243386} = .8558$$

b. We would expect most of the cocoon temperatures to be within $2s$ of the least squares line. This is:

$$2s = 2(.8558) = 1.7116$$

c. To determine if the outside air temperature provides information for predicting the woolly-bear caterpillar cocoon temperature, we test:

H_0: $\beta_1 = 0$
H_a: $\beta_1 \neq 0$

The test statistic is $t = \dfrac{\hat{\beta}_1 - 0}{s_{\hat{\beta}_1}} = \dfrac{1.201 - 0}{\dfrac{.8558}{\sqrt{83.3425}}} = 12.81$

The rejection region requires $\alpha/2 = .05/2 = .025$ in each tail of the t-distribution with df $= n - 2 = 12 - 2 = 10$. From Table IV, Appendix A, $t_{.025} = 2.228$. The rejection region is $t < -2.228$ or $t > 2.228$.

Since the observed value of the test statistic falls in the rejection region ($t = 12.81 > 2.228$), H_0 is rejected. There is sufficient evidence to indicate that the outside air temperature provides information for predicting the woolly-bear caterpillar cocoon temperature at $\alpha = .05$.

d. Since $12.81 > 4.587 = t_{.0005}$ with 10 df, p-value $< 2(.0005) = .001$

9.45 a. $r = 1$ implies x and y are perfectly, positively related.

b. $r = -1$ implies x and y are perfectly, negatively related.

c. $r = 0$ implies x and y are not related.

d. $r = .90$ implies x and y are positively related. Since r is close to 1, the strength of the relationship is very high.

e. $r = .10$ implies x and y are positively related. Since r is close to 0, the relationship is fairly weak.

f. $r = -.88$ implies x and y are negatively related. Since r is close to -1, the relationship is fairly strong.

9.47 a. Some preliminary calculations are:

$$\sum x = 0 \qquad \sum x^2 = 10 \qquad \sum xy = 20$$
$$\sum y = 12 \qquad \sum y^2 = 70$$

$$SS_{xy} = \sum xy - \frac{\sum x \sum y}{n} = 20 - \frac{0(12)}{5} = 20$$

$$SS_{xx} = \sum x^2 - \frac{(\sum x)^2}{n} = 10 - \frac{0^2}{5} = 10$$

$$SS_{yy} = \sum y^2 - \frac{(\sum y)^2}{n} = 70 - \frac{12^2}{5} = 41.2$$

$$r = \frac{SS_{xy}}{\sqrt{SS_{xx}SS_{yy}}} = \frac{20}{\sqrt{10(41.2)}} = .9853$$

$$r^2 = .9853^2 = .9709$$

Since $r = .9853$, there is a very strong positive linear relationship between x and y.

Since $r^2 = .9709$, 97.09% of the total sample variability around $\bar{y}$ is explained by the linear relationship between x and y.

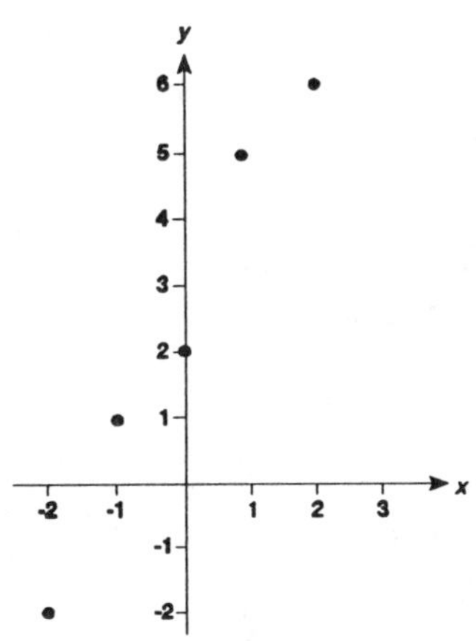

b. Some preliminary calculations are:

$$\sum x = 0 \qquad \sum x^2 = 10 \qquad \sum xy = -15$$
$$\sum y = 16 \qquad \sum y^2 = 74$$

$$SS_{xy} = \sum xy - \frac{\sum x \sum y}{n} = -15 - \frac{0(16)}{5} = -15$$

$$SS_{xx} = \sum x^2 - \frac{\left(\sum x\right)^2}{n} = 10 - \frac{0^2}{5} = 10$$

$$SS_{yy} = \sum y^2 - \frac{\left(\sum y\right)^2}{n} = 74 - \frac{16^2}{5} = 22.8$$

$$r = \frac{SS_{xy}}{\sqrt{SS_{xx}SS_{yy}}} = \frac{-15}{\sqrt{10(22.8)}} = -.9934$$

$$r^2 = (-.9934)^2 = .9868$$

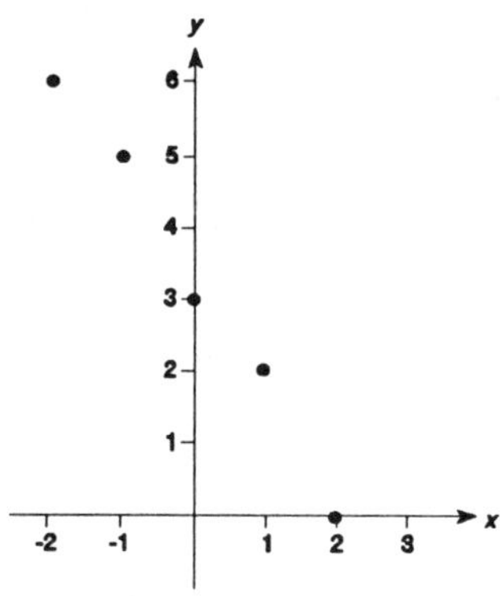

Since $r = -.9934$, there is a very strong negative linear relationship between x and y.

Since $r^2 = .9868$, 98.68% of the total sample variability around $\bar{y}$ is explained by the linear relationship between x and y.

c. Some preliminary calculations are:

$$\sum x = 18 \qquad \sum x^2 = 52 \qquad \sum xy = 36$$
$$\sum y = 14 \qquad \sum y^2 = 32$$

$$SS_{xy} = \sum xy - \frac{\sum x \sum y}{n} = 36 - \frac{18(14)}{7} = 0$$

$$SS_{xx} = \sum x^2 - \frac{\left(\sum x\right)^2}{n} = 52 - \frac{18^2}{7} = 5.71428571$$

$$SS_{yy} = \sum y^2 - \frac{\left(\sum y\right)^2}{n} = 32 - \frac{14^2}{7} = 4$$

$$r = \frac{SS_{xy}}{\sqrt{SS_{xx}SS_{yy}}} = \frac{0}{\sqrt{5.71428571(4)}} = 0$$

$$r^2 = 0^2 = 0$$

Since $r = 0$, this implies that x and y are not related.

Since $r^2 = 0$, 0% of the total sample variability around $\bar{y}$ is explained by the linear relationship between x and y.

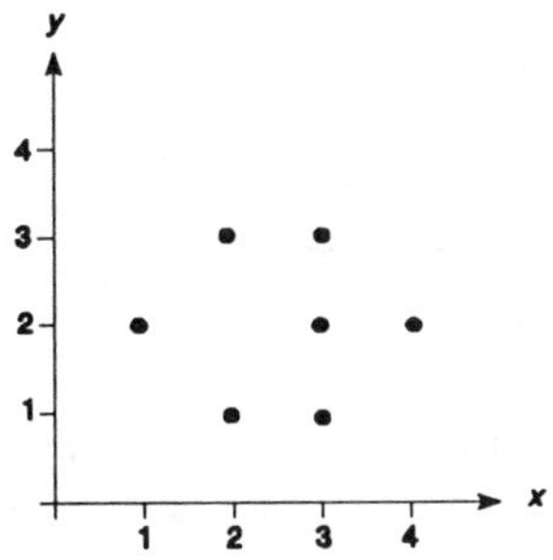

d. Some preliminary calculations are:

$$\sum x = 15 \qquad \sum x^2 = 71 \qquad \sum xy = 12$$
$$\sum y = 4 \qquad \sum y^2 = 6$$
$$SS_{xy} = \sum xy - \frac{\sum x \sum y}{n} = 12 - \frac{15(4)}{5} = 0$$
$$SS_{xx} = \sum x^2 - \frac{(\sum x)^2}{n} = 71 - \frac{15^2}{5} = 26$$
$$SS_{yy} = \sum y^2 - \frac{(\sum y)^2}{n} = 6 - \frac{4^2}{5} = 2.8$$
$$r = \frac{SS_{xy}}{\sqrt{SS_{xx}SS_{yy}}} = \frac{0}{\sqrt{26(2.8)}} = 0$$
$$r^2 = 0^2 = 0$$

Since $r = 0$, this implies that x and y are not related.

Since $r^2 = 0$, 0% of the total sample variability around $\bar{y}$ is explained by the linear relationship between x and y.

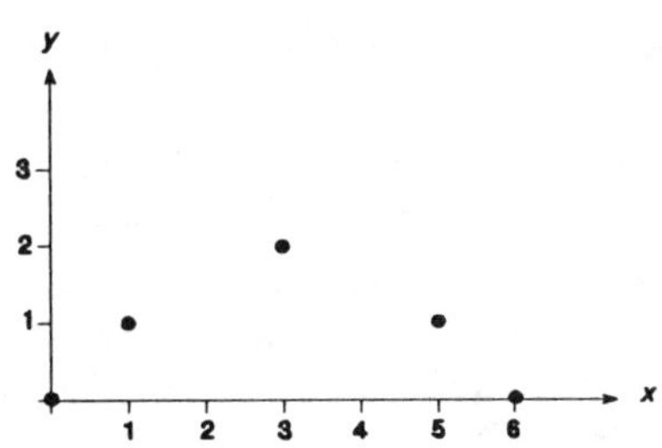

9.49 a. $\hat{\beta}_0 = 175.4$. This is the estimated value of the y-intercept. No other information is available.

$\hat{\beta}_1 = 133.2$. For each unit increase in the Index of Difficulty, the mean increase in the time is estimated to be 133.2.

b. $r = .951$. There is a fairly strong positive linear relationship between the Index of Difficulty and the time required to move and select a target.

c. To determine if Fitt's Law model is statistically adequate for predicting performance time, we test:

H_0: $\beta_1 = 0$
H_a: $\beta_1 \neq 0$

The test statistic is $t = \dfrac{r}{\sqrt{(1 - r^2)/(n - 2)}} = \dfrac{.951}{\sqrt{(1 - .951^2)/(160 - 2)}} = 38.66$

The rejection region requires $\alpha/2 = .05/2 = .025$ in each tail of the t-distribution with df $= n - 2 = 160 - 2 = 158$. From Table IV, Appendix A, $t_{.025} \approx 1.960$. The rejection region is $t < -1.96$ or $t > 1.96$.

Since the observed value of the test statistic falls in the rejection region ($t = 38.66 > 1.96$), H_0 is rejected. There is sufficient evidence to indicate that Fitt's Law model is statistically adequate for predicting performance time at $\alpha = .05$.

d. $r^2 = .951^2 = .904$. 90.4% of the total sample variability around the mean performance time is explained by the linear relationship between performance time and the Index of Difficulty.

9.51 a. $r = .26$. Because this value is fairly close to 0, there is a weak positive linear relationship between a daughter's loneliness score and a mother's loneliness score.

b. $r = .19$. Because this value is fairly close to 0, there is a weak positive linear relationship between a daughter's loneliness score and a father's loneliness score.

c. $r = .14$. Because this value is fairly close to 0, there is a weak positive linear relationship between a daughter's loneliness score and a mother's self-esteem score.

d. $r = .01$. Because this value is extremely close to 0, there is an extremely weak positive linear relationship between a daughter's loneliness score and a father's assertiveness score. Because this value is so close to 0, these two variables are probably not linearly related.

e. (a) $r^2 = .26^2 = .0676$. 6.76% of the total sample variability around the sample mean daughter's loneliness score is explained by the linear relationship between a daughter's loneliness score and a mother's loneliness score.
(b) $r^2 = .19^2 = .0361$. 3.61% of the total sample variability around the sample mean daughter's loneliness score is explained by the linear relationship between a daughter's loneliness score and a father's loneliness score.
(c) $r^2 = .14^2 = .0196$. 1.96% of the total sample variability around the sample mean daughter's loneliness score is explained by the linear relationship between a daughter's loneliness score and a mother's self-esteem score.
(d) $r^2 = .01^2 = .0001$. 0.01% of the total sample variability around the sample mean daughter's loneliness score is explained by the linear relationship between a daughter's loneliness score and a father's assertiveness score.

9.53 a. A straight line model is $y = \beta_0 + \beta_1 x + \epsilon$.

b. The researcher hypothesized that therapists with more years of formal dance training will report a higher perceived success rate in cotherapy relationships. This indicates that $\beta_1 > 0$.

c. $r = -.26$. Because this value is fairly close to 0, there is a weak negative linear relationship between years of formal training and reported success rate.

d. To determine if there is a positive linear relationship between years of formal training and reported success rate, we test:

$$H_0\text{: } \beta_1 = 0$$
$$H_0\text{: } \beta_1 > 0$$

The test statistic is $t = \dfrac{r}{\sqrt{(1 - r^2)/(n - 2)}} = \dfrac{-.26}{\sqrt{(1 - (-.26^2))/(136 - 2)}} = -3.12$

The rejection region requires $\alpha = .05$ in the upper tail of the t-distribution with df $= n - 2 = 136 - 2 = 134$. From Table IV, Appendix A, $t_{.05} \approx 1.645$. The rejection region is $t > 1.645$.

Since the observed value of the test statistic does not fall in the rejection region ($t = -8.66 \not> 1.645$), H_0 is not rejected. There is insufficient evidence to indicate that there is a positive linear relationship between years of formal training and perceived success rates at $\alpha = .05$.

9.55 a. $r = .679$. Because this value is near .5, there is a moderate positive linear relationship between the fraction of documents retrieved using Medline and the number of terms in the search query.

b. $r^2 = .679^2 = .461$. 46.1% of the total sample variability around the sample mean fraction of documents retrieved using Medline is explained by the linear relationship between the fraction of documents retrieved using Medline and the number of terms in the search query.

9.57 a. To determine whether the number of leisure activities and GPA are linearly related, we test:

$$H_0\text{: } \rho = 0$$
$$H_a\text{: } \rho \neq 0$$

b. The p-value $= .0512$. This implies that H_0 will not be rejected for $\alpha = .05$. There is insufficient evidence to indicate that the number of leisure activities and GPA are linearly related for $\alpha = .05$. For $\alpha > .0512$, H_0 will be rejected. There is sufficient evidence to indicate that the number of leisure activities and GPA are linearly related for $\alpha > .0512$.

9.59 a,b.

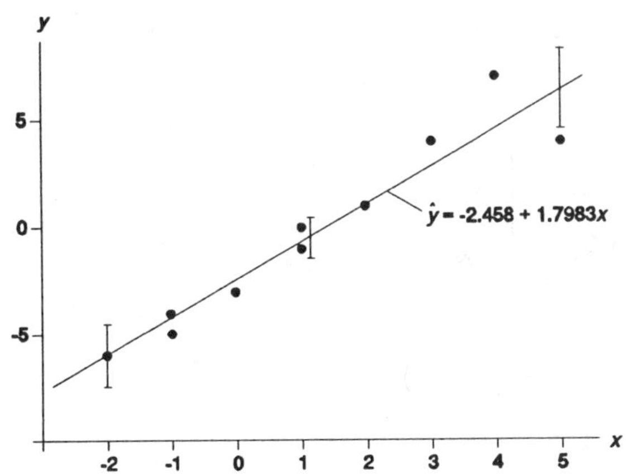

c. Some preliminary calculations are:

$$SSE = SS_{yy} - \hat{\beta}_1 SS_{xy} = 168.1 - 1.798319328(85.6) = 14.1638655$$

$$s^2 = \frac{SSE}{n-2} = \frac{14.1638655}{10-2} = 1.770483188 \qquad s = \sqrt{1.770483188} = 1.3306$$

$$\bar{x} = \frac{\sum x}{n} = \frac{12}{10} = 1.2$$

For $x_p = 5$, $\hat{y} = -2.458 + 1.7983(5) = 6.5335$

For confidence coefficient .95, $\alpha = 1 - .95 = .05$ and $\alpha = .05/2 = .025$. From Table IV, Appendix A, $t_{.025} = 2.306$ with df $= n - 2 = 10 - 2 = 8$. The 95% confidence interval is:

$$\hat{y} \pm t_{\alpha/2} s \sqrt{\frac{1}{n} + \frac{(x_p - \bar{x})^2}{SS_{xx}}} \Rightarrow 6.5335 \pm 2.306(1.3306)\sqrt{\frac{1}{10} + \frac{(5 - 1.2)^2}{47.6}}$$

$$\Rightarrow 6.5335 \pm 1.9487 \Rightarrow (4.5848, 8.4822)$$

d. For $x_p = 1.2$, $\hat{y} = -2.458 + 1.7983(1.2) = -.3$. The 95% confidence interval is:

$$-.3 \pm 2.306(1.3306)\sqrt{\frac{1}{10} + \frac{(1.2 - 1.2)^2}{47.6}} \Rightarrow -.3 \pm .9703 \Rightarrow (-1.2703, .6703)$$

For $x_p = -2$, $\hat{y} = -2.458 + 1.7983(-2) = -6.0546$. The 95% confidence interval is:

$$-6.0546 \pm 2.306(1.3306)\sqrt{\frac{1}{10} + \frac{(-2 - 1.2)^2}{47.6}} \Rightarrow -6.0546 \pm 1.7225$$

$$\Rightarrow (-7.7771, -4.3321)$$

e. The width of the confidence interval when $x = 1.2$ is smaller than when $x = -2$ or $x = 5$. The widths of the intervals change because of the term $\frac{(x_p - \bar{x})^2}{SS_{xx}}$. If $x_p = \bar{x}$, the term is 0. As x_p moves further away from $\bar{x}$, the width of the interval increases.

9.61 a, b. The scattergram is:

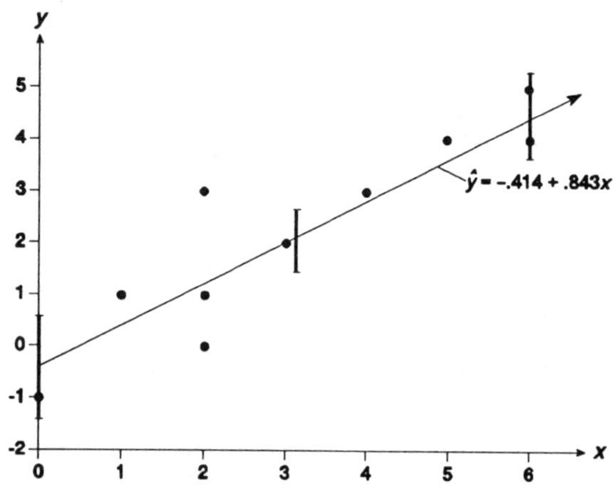

c. $SSE = SS_{yy} - \hat{\beta}_1 SS_{xy} = 33.6 - .84318766(32.8) = 5.94344473$

$s^2 = \dfrac{SSE}{n-2} = \dfrac{5.94344473}{10-2} = .742930591$ $\qquad s = \sqrt{.742930591} = .8619$

$\bar{x} = \dfrac{31}{10} = 3.1$

The form of the confidence interval is $\hat{y} \pm t_{\alpha/2} s \sqrt{\dfrac{1}{n} + \dfrac{(x_p - \bar{x})^2}{SS_{xx}}}$

For $x_p = 6$, $\hat{y} = -.414 + .843(6) = 4.644$

For confidence coefficient .95, $\alpha = .05$ and $\alpha/2 = .025$. From Table IV, Appendix A, with df $= n - 2 = 10 - 2 = 8$, $t_{.025} = 2.306$. The confidence interval is:

$$4.644 \pm 2.306(.8619)\sqrt{\frac{1}{10} + \frac{(6 - 3.1)^2}{38.9}} \Rightarrow 4.644 \pm 1.118 \Rightarrow (3.526, 5.762)$$

d. For $x_p = 3.2$, $\hat{y} = -.414 + .843(3.2) = 2.284$

The confidence interval is:

$$2.284 \pm 2.306(.8619)\sqrt{\frac{1}{10} + \frac{(3.2 - 3.1)^2}{38.9}} \Rightarrow 2.284 \pm .629 \Rightarrow (1.655, 2.913)$$

For $x_p = 0$, $\hat{y} = -.414 + .843(0) = -.414$

The confidence interval is:

$$-.414 \pm 2.306(.8619)\sqrt{\frac{1}{10} + \frac{(0 - 3.1)^2}{38.9}} \Rightarrow -.414 \pm 1.717 \Rightarrow (-1.585, .757)$$

e. The width of the confidence interval for the mean value of y depends on the distance x_p is from $\bar{x}$. The width of the interval for $x_p = 3.2$ is the smallest because 3.2 is the closest to $\bar{x} = 3.1$. The width of the interval for $x_p = 0$ is the widest because 0 is the farthest from $\bar{x} = 3.1$.

9.63 a. The prediction interval is (4.5504, 6.3575). We are 95% confident that the best jumping distance of a novice long jumper will be between 4.5504 and 6.3575 meters when the average takeoff error is .20 meters.

b. The confidence interval is (5.2191, 5.6888). We are 95% confident that the mean best jumping distance of all novice long jumpers will be between 5.2191 and 5.6888 meters when the average takeoff error is .20 meters.

c. From Exercise 9.40, the test to determine if the model was adequate, the conclusions was "Do not reject H_0." There is insufficient evidence to indicate the model is adequate. Thus, the model should not be used to predict the best jumping distance.

9.65 a. Some preliminary calculations are:

$$\sum x_i = 16.0 \qquad \sum y_i = 101 \qquad n = 6$$
$$\sum x_i^2 = 46.5 \qquad \sum y_i^2 = 1815 \qquad \sum x_i y_i = 289.5$$

$$SS_{xy} = \sum xy - \frac{\sum x \sum y}{n} = 289.5 - \frac{(16.0)(101)}{6} = 20.166667$$

$$SS_{xx} = \sum x^2 - \frac{(\sum x)^2}{n} = 46.5 - \frac{(16.0^2)}{6} = 3.833333$$

$$SS_{yy} = \sum y^2 - \frac{(\sum y)^2}{n} = 1815 - \frac{(101^2)}{6} = 114.83333$$

$$\hat{\beta}_1 = \frac{SS_{xy}}{SS_{xx}} = \frac{20.166667}{3.8333333} = 5.260869566 \approx 5.2609$$

$$\hat{\beta}_0 = \bar{y} - \hat{\beta}_1 \bar{x} = \frac{101}{6} - (5.260869566)\left(\frac{16.0}{6}\right) = 2.804347825 \approx 2.804$$

$$s^2 = \frac{SS_{yy} - \hat{\beta}_1 SS_{xy}}{n - 2} = \frac{114.83333 - 5.260869566(20.166667)}{6 - 2} = 2.184782616$$

The least squares equation is $\hat{y} = 2.804 + 5.2609x$.

To determine if there is sufficient evidence to indicate a linear relationship between drug dosage and change in pulse rate, we test:

H_0: $\beta_1 = 0$
H_a: $\beta_1 \neq 0$

The test statistic is $t = \dfrac{\hat{\beta}_1 - 0}{s_{\hat{\beta}_1}} = \dfrac{5.2609}{\dfrac{\sqrt{2.184782616}}{\sqrt{3.8333333}}} = 6.97$

The rejection region requires $\alpha/2 = .10/2 = .05$ in each tail of the t-distribution with df $= n - 2 = 6 - 2 = 4$. From Table IV, Appendix A, $t_{.05} = 2.132$. The rejection region is $t < -2.132$ or $t > 2.132$.

Since the observed value of the test statistic falls in the rejection region ($t = 6.97 > 2.132$), H_0 is rejected. There is sufficient evidence to indicate that there is a linear relationship between drug dosage and change in pulse rate at $\alpha = .10$.

b. The form of the confidence interval for β_1 is $\hat{\beta}_1 \pm t_{\alpha/2} s_{\hat{\beta}_1}$

For confidence coefficient .95, $\alpha = .05$ and $\alpha/2 = .025$. From Table IV, Appendix A, with df $= n - 2 = 6 - 2 = 4$, $t_{.025} = 2.776$. The confidence interval is:

$$5.261 \pm 2.776\left(\frac{\sqrt{2.18478}}{\sqrt{3.8333333}}\right) \Rightarrow 5.261 \pm 2.096 \Rightarrow (3.165, 7.357)$$

c. The form of the confidence interval is $\hat{y} \pm t_{\alpha/2} s \sqrt{\dfrac{1}{n} + \dfrac{(x_p - \bar{x})^2}{SS_{xx}}}$

For $x_p = 3.5$, $\hat{y} = 2.804 + 5.2609(3.5) = 21.22$

For confidence coefficient .99, $\alpha = .01$ and $\alpha/2 = .005$. From Table IV, Appendix A, with df $= n - 2 = 6 - 2 = 4$, $t_{.005} = 4.604$. The confidence interval is:

$$21.22 \pm 4.604\sqrt{2.18478}\sqrt{\frac{1}{6} + \frac{\left(3.5 - \frac{16}{6}\right)^2}{3.8333333}} \Rightarrow 21.22 \pm 4.01 \Rightarrow (17.21, 25.23)$$

9.67 a. To determine if the average hourly wage rate contributes information to predict quit rates, we test:

H_0: $\beta_1 = 0$
H_a: $\beta_1 \neq 0$

The test statistic is $t = \dfrac{\hat{\beta}_1 - 0}{s_{\hat{\beta}_1}} = \dfrac{-.3466 - 0}{.05866} = -5.91$

The rejection region requires $\alpha/2 = .05/2 = .025$ in each tail of the t-distribution with df $= n - 2 = 15 - 2 = 13$. From Table IV, Appendix A, $t_{.025} = 2.160$. The rejection region is $t < -2.160$ or $t > 2.160$.

Since the observed value of the test statistic falls in the rejection region ($t = -5.91 < -2.160$), H_0 is rejected. There is sufficient evidence to indicate that the average hourly wage rate contributes information to predict quit ratio at $\alpha = .05$.

Since the slope is negative ($\hat{\beta}_1 = -.3466$), the model suggests that x and y have a negative relationship. As the average hourly wage rate increases, the quit rate tends to decrease.

b. From the printout, the 95% prediction interval is (.656, 2.829). We are 95% confident that the quit rate in an industry with an average hourly wage of $9.00 is between .656 and 2.829.

c. From the printout, the 95% confidence interval is (1.467, 2.018). We are 95% confident that the mean quit rate of all industries with an average hourly wage of $9.00 is between 1.467 and 2.018.

9.69 a. From Exercise 9.31, $SS_{xx} = 3000$ and $\bar{x} = 50$.

Also, for Brand A, $s = 1.211$; for Brand B, $s = .610$.

For Brand A, $\hat{y} = 6.62 - .0727(45) = 3.349$, while for Brand B, $\hat{y} = 9.31 - .1077(45) = 4.464$.

The degrees of freedom for both brands is $n - 2 = 15 - 2 = 13$. For confidence coefficient .90, (i.e., for all parts of this question), $\alpha = .10$ and $\alpha/2 = .05$. From Table IV, Appendix A, with df $= 13$, $t_{.05} = 1.771$.

The form of both confidence intervals is $\hat{y} \pm t_{\alpha/2}s\sqrt{\frac{1}{n} + \frac{(x_p - \bar{x})^2}{SS_{xx}}}$

For Brand A, we obtain:

$$3.349 \pm 1.771(1.211)\sqrt{\frac{1}{15} + \frac{(45 - 50)^2}{3000}} \Rightarrow 3.349 \pm .587 \Rightarrow (2.762, 3.936)$$

For Brand B, we obtain:

$$4.464 \pm 1.771(.610)\sqrt{\frac{1}{15} + \frac{(45 - 50)^2}{3000}} \Rightarrow 4.464 \pm .296 \Rightarrow (4.168, 4.760)$$

The first interval is wider, caused by the larger value of s.

b. The form of both prediction intervals is $\hat{y} \pm t_{\alpha/2}s\sqrt{1 + \frac{1}{n} + \frac{(x_p - \bar{x})^2}{SS_{xx}}}$

For Brand A, we obtain:

$$3.349 \pm 1.771(1.211)\sqrt{1 + \frac{1}{15} + \frac{(45 - 50)^2}{3000}} \Rightarrow 3.349 \pm 2.224 \Rightarrow (1.125, 5.573)$$

For Brand B, we obtain:

$$4.464 \pm 1.771(.610)\sqrt{1 + \frac{1}{15} + \frac{(45 - 50)^2}{3000}} \Rightarrow 4.464 \pm 1.120 \Rightarrow (3.344, 5.584)$$

Again, the first interval is wider, caused by the larger value of s. Each of these intervals is wider than its counterpart from part **a**, since, for the same x, a prediction interval for an individual y is always wider than a confidence interval for the mean of y. This is due to an individual observation having a greater variance than the variance of the mean of a set of observations.

c. To obtain a confidence interval for the life of a brand A cutting tool that is operated at 100 meters per minute, we use:

$$\hat{y} \pm t_{\alpha/2}s\sqrt{1 + \frac{1}{n} + \frac{(x_p - \bar{x})^2}{SS_{xx}}}$$

For $x = 100$, $\hat{y} = 6.62 - .0727(100) = -.65$.

The degrees of freedom are $n - 2 = 15 - 2 = 13$. For confidence coefficient .95, $\alpha = .05$ and $\alpha/2 = .025$. From Table IV, Appendix A, with df = 13, $t_{.025} = 2.160$.

Here, we obtain:

$$-.65 \pm 2.160(1.211)\sqrt{1 + \frac{1}{15} + \frac{(100 - 50)^2}{3000}} \Rightarrow -.65 \pm 3.606 \Rightarrow (-4.256, 2.956)$$

The additional assumption would be that the straight line model fits the data well for the x's actually observed all the way up to the value under consideration, 100. Clearly from the estimated value of $-.65$, this is not true (usually, negative "useful lives" are not found).

9.71 Since there are no ties, we will use the shortcut formula.

a. Some preliminary calculations are:

x Rank (u_i)	y Rank (v_i)	$d_i = u_i - v_i$	d_i^2
3	2	1	1
5	4	1	1
2	5	−3	9
1	1	0	0
4	3	1	1
		Total =	12

$$r_s = 1 - \frac{6\sum d_i^2}{n(n^2 - 1)} = 1 - \frac{6(12)}{5(5^2 - 1)} = 1 - .6 = .4$$

b.

x Rank (u_i)	y Rank (v_i)	$d_i = u_i - v_i$	d_i^2
2	3	−1	1
3	4	−1	1
4	2	2	4
5	1	4	16
1	5	−4	16
		Total =	38

$$r_s = 1 - \frac{6\sum d_i^2}{n(n^2 - 1)} = 1 - \frac{6(38)}{5(5^2 - 1)} = 1 - 1.9 = -.9$$

c.

x Rank (u_i)	y Rank (v_i)	$d_i = u_i - v_i$	d_i^2
1	2	−1	1
4	1	3	9
2	3	−1	1
3	4	−1	1
		Total =	12

$$r_s = 1 - \frac{6\sum d_i^2}{n(n^2 - 1)} = 1 - \frac{6(12)}{4(4^2 - 1)} = 1 - 1.2 = -.2$$

d.

x Rank (u_i)	y Rank (v_i)	$d_i = u_i - v_i$	d_i^2
2	1	1	1
5	3	2	4
4	5	−1	1
3	2	1	1
1	4	−3	9
		Total =	16

$$r_s = 1 - \frac{6\sum d_i^2}{n(n^2 - 1)} = 1 - \frac{6(16)}{5(5^2 - 1)} = 1 - .8 = .2$$

9.73 a. The Spearman's rank correlation between MODA and face-matching scores is $r_s = .48$. There is a moderately weak, positive relationship between MODA and face-matching scores.

b. To determine if there is a positive correlation between MODA scores and face-matching scores for Alzheimer's patients, we test:

H_0: $\rho_s = 0$
H_a: $\rho_s > 0$

The test statistic is $r_s = .48$

Reject H_0 if $r_s > r_{s,\alpha}$ where $\alpha = .05$ and $n = 30$:

Reject H_0 if $r_s > .305$ (using Table XII, Appendix A)

Since the observed value of the test statistic falls in the rejection region ($r_s = .48 > .305$), H_0 is rejected. There is sufficient evidence to indicate that there is a positive correlation between MODA scores and face-matching scores for Alzheimer's patients at $\alpha = .05$.

9.75 a. The correlation between a variable and itself will always be 1.00. Therefore, the Spearman's rank correlations on the diagonals are always 1.00.

b. The Spearman's rank correlation between EDSS and ARM is −.439. There is a moderately weak, negative relationship between the Expanded Disability Status Scale and the peak oxygen uptake during arm cranking exercise.

The Spearman's rank correlation between EDSS and LEG/ARM is −.655. There is a moderately strong, negative relationship between the Expanded Disability Status Scale and the peak oxygen uptake during leg cycling and arm cranking exercise.

The Spearman's rank correlation between EDSS and LEG is −.512. There is a moderate, negative relationship between the Expanded Disability Status Scale and the peak oxygen uptake during leg cycling exercise.

The Spearman's rank correlation between ARM and LEG/ARM is .634. There is a moderately strong, positive relationship between the peak oxygen uptake during arm cranking exercise and the peak oxygen uptake during combined leg cycling and arm cranking exercise.

The Spearman's rank correlation between ARM and LEG is .722. There is a moderately strong, positive relationship between the peak oxygen uptake during arm cranking exercise and the peak oxygen uptake during leg cycling exercise.

The Spearman's rank correlation between LEG/ARM and LEG is .890. There is a strong, positive relationship between the peak oxygen uptake during combined leg cycling and arm cranking exercise and peak oxygen uptake during leg cycling exercise.

c. To determine if EDSS and peak oxygen uptake for arm cranking exercise are negatively correlated, we test:

H_0: $\rho_s = 0$
H_a: $\rho_s < 0$

The test statistic is $r_s = -.439$.

Reject H_0 is $r_s < -r_{s,\alpha}$ where $\alpha = .01$ and $n = 10$:

Reject H_0 if $r_s < -.745$ (using Table XII, Appendix A)

Since the observed value of the test statistic does not fall in the rejection region ($r_s = -.439 \nless -.745$), H_0 is not rejected. There is insufficient evidence to indicate that there is a negative correlation between EDSS and peak oxygen uptake for arm cranking exercise at $\alpha = .01$.

To determine if EDSS and peak oxygen uptake for combined leg cycling and arm cranking exercise are negatively correlated, we test:

H_0: $\rho_s = 0$
H_a: $\rho_s < 0$

The test statistic is $r_s = -.655$.

Reject H_0 is $r_s < -r_{s,\alpha}$ where $\alpha = .01$ and $n = 10$:

Reject H_0 if $r_s < -.745$ (using Table XII, Appendix A)

Since the observed value of the test statistic does not fall in the rejection region ($r_s = -.655 \nless -.745$), H_0 is not rejected. There is insufficient evidence to indicate that there is a negative correlation between EDSS and peak oxygen uptake for combined leg cycling and arm cranking exercise at $\alpha = .01$.

To determine if EDSS and peak oxygen uptake for leg cycling exercise are negatively correlated, we test:

H_0: $\rho_s = 0$
H_a: $\rho_s < 0$

The test statistic is $r_s = -.512$.

Reject H_0 is $r_s < -r_{s,\alpha}$ where $\alpha = .01$ and $n = 10$:

Reject H_0 if $r_s < -.745$ (using Table XII, Appendix A)

Since the observed value of the test statistic does not fall in the rejection region ($r_s = -.512 \nless -.745$), H_0 is not rejected. There is insufficient evidence to indicate that there is a negative correlation between EDSS and peak oxygen uptake for leg cycling exercise at $\alpha = .01$.

9.77 Some preliminary calculations are:

Student	Deception Score	Rank	Percentage Increase in Eye Pupil Dilation	Rank	d_i	d_i^2
1	87	6	10	6	0	0
2	63	5	6	4	1	1
3	95	8	11	7	1	1
4	50	3	7	5	−2	4
5	43	2	0	1	1	1
6	89	7	15	8	−1	1
7	33	1	4	2	−1	1
8	55	4	5	3	1	1
						$\sum d_i^2 = 10$

To determine if the percentage increase in eye pupil size is positively correlated to deception score, we test:

H_0: $\rho_s = 0$
H_a: $\rho_s > 0$

The test statistic is $r_s = 1 - \dfrac{6\sum d_i^2}{n(n^2 - 1)} = 1 - \dfrac{6(10)}{8(8^2 - 1)} = 1 - .119 = .881$

Reject H_0 if $r_s > r_{s,\alpha}$ where $\alpha = .05$ and $n = 8$:

Reject H_0 if $r_s > .643$ (from Table XII, Appendix A).

Since the observed value of the test statistic falls in the rejection region ($r_s = .881 > .643$), reject H_0. There is sufficient evidence to indicate increase in eye pupil dilation is positively correlated with deception score at $\alpha = .05$.

9.79 Some preliminary calculations are:

Category	Crisis Intervention Rating	Rank, u_i	Clarity Rating	Rank, v_i	d_i	d_i^2
Psychosis	1.31	3	1.33	3	0	0
Drug/alcohol abuse	1.33	4	1.29	1	3	9
Depression/anxiety	1.48	5	1.59	4	1	1
Emphasis on acuteness	1.76	6	2.50	8	−2	4
Insistence on "short-term" response	2.48	7	3.22	9	−2	4
Suicide	1.13	2	1.32	2	0	0
Family problems	2.59	8	2.30	6	2	4
Violence/harm	1.06	1	1.86	5	−4	16
Miscellaneous	2.60	9	2.33	7	2	4
Nondefinition	3.57	10	3.57	10	0	0
						$\sum d_i^2 = 42$

To determine if there is a positive relationship between the mean crisis intervention and mean clarity ratings, we test:

H_0: $\rho_s = 0$
H_a: $\rho_s > 0$

The test statistic is $r_s = 1 - \dfrac{6\sum d_i^2}{n(n^2 - 1)} = 1 - \dfrac{6(42)}{10(10^2 - 1)} = .255$

Reject H_0 if $r_s > r_{s,\alpha}$ where $\alpha = .05$ and $n = 10$:

Reject H_0 if $r_s > .564$ (using Table XII, Appendix A).

Since the observed value of the test statistic does not fall in the rejection region ($r_s = .255 \not> .564$), H_0 is not rejected. There is insufficient evidence to indicate that there is a positive relationship between the mean crisis intervention and mean clarity ratings at $\alpha = .05$.

9.81 a.

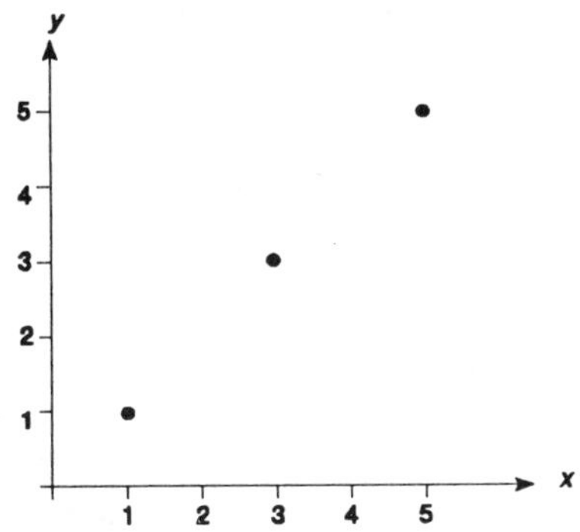

b. One possible line is $\hat{y} = x$.

x	y	$\hat{y}$	$y - \hat{y}$
1	1	1	0
3	3	3	0
5	5	5	0
			0

For this example $\sum (y - \hat{y}) = 0$

A second possible line is $\hat{y} = 3$.

x	y	$\hat{y}$	$y - \hat{y}$
1	1	3	−2
3	3	3	0
5	5	3	2
			0

For this example $\sum (y - \hat{y}) = 0$

c. Some preliminary calculations are:

$\sum x_i = 9 \qquad \sum x_i^2 = 35 \qquad \sum x_i y_i = 35$

$\sum y_i = 9 \qquad \sum y_i^2 = 35$

$$SS_{xy} = \sum x_i y_i - \frac{\sum x_i \sum y_i}{n} = 35 - \frac{9(9)}{3} = 8$$

$$SS_{xx} = \sum x_i^2 - \frac{\left(\sum x_i\right)^2}{n} = 35 - \frac{9^2}{3} = 8$$

$$SS_{yy} = \sum y_i^2 - \frac{\left(\sum y_i\right)^2}{n} = 35 - \frac{9^2}{3} = 8$$

$$\hat{\beta}_1 = \frac{SS_{xy}}{SS_{xx}} = \frac{8}{8} = 1 \qquad \hat{\beta}_0 = \bar{y} - \hat{\beta}_1\bar{x} = \frac{9}{3} - 1\left(\frac{9}{3}\right) = 0$$

The least squares line is $\hat{y} = 0 + 1x = x$.

d. For $\hat{y} = x$, SSE $= SS_{yy} - \hat{\beta}_1 SS_{xy} = 8 - 1(8) = 0$

For $\hat{y} = 3$, SSE $= \sum(y_i - \hat{y}_i)^2 = (1 - 3)^2 + (3 - 3)^2 + (5 - 3)^2 = 8$

The least squares line has the smallest SSE of all possible lines.

9.83 a. From the printout, the least squares equation is $\hat{y} = 40.7842 + .7656x$.

b.

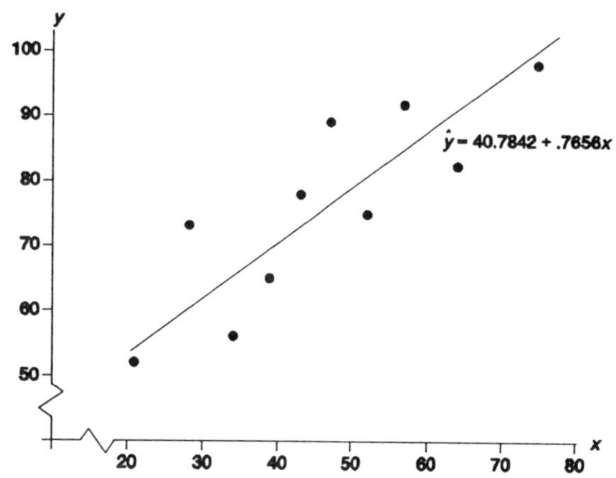

c. To determine if there is sufficient evidence to indicate that a positive correlation exists between verbal score and final grade, we test:

H_0: $\beta_1 = 0$
H_a: $\beta_1 > 0$

The test statistic is $t = \frac{\hat{\beta}_1 - 0}{s_{\hat{\beta}_1}} = \frac{.765562}{.714985} = 4.375$

The rejection region requires $\alpha = .01$ in the upper tail of the t-distribution with df $= n - 2 = 10 - 2 = 8$. From Table IV, Appendix A, $t_{.01} = 2.896$. The rejection region is $t > 2.896$.

Since the observed value of the test statistic falls in the rejection region ($t = 4.375 > 2.896$), H_0 is rejected. There is sufficient evidence to indicate that a positive correlation exists between verbal score and final grade at $\alpha = .01$.

d. The form of the confidence interval for β_1 is $\hat{\beta}_1 \pm t_{\alpha/2}s_{\hat{\beta}_1}$

For confidence coefficient .95, $\alpha = .05$ and $\alpha/2 = .025$. From Table IV, Appendix A, with df $= n - 2 = 10 - 2 = 8$, $t_{.025} = 2.306$.

The confidence interval is $.7656 \pm 2.306(.174985) \Rightarrow .7656 \pm .4035 \Rightarrow (.3621, 1.1691)$.

e. The form of the prediction interval is $\hat{y} \pm t_{\alpha/2}s\sqrt{1 + \frac{1}{n} + \frac{(x_p - \bar{x})^2}{SS_{xx}}}$

For $x_p = 50$, $\hat{y} = 40.7842 + .7656(50) = 79.06$.

For confidence coefficient .90, $\alpha = .10$ and $\alpha/2 = .05$. From Table IV, Appendix A, with df $= n - 2 = 10 - 2 = 8$, $t_{.05} = 1.860$.

The prediction interval is:

$$79.06 \pm 1.860\sqrt{75.753233}\sqrt{1 + \frac{1}{10} + \frac{(50 - 46)^2}{2474}} \Rightarrow 79.06 \pm 17.03$$
$$\Rightarrow (62.03, 96.09)$$

f. The form of the confidence interval is $\hat{y} \pm t_{\alpha/2}s\sqrt{\frac{1}{n} + \frac{(x_p - \bar{x})^2}{SS_{xx}}}$

For $x_p = 35$, $\hat{y} = 40.7842 + .7656(35) = 67.58$

For confidence coefficient .95, $\alpha = .05$ and $\alpha/2 = .025$. From Table IV, Appendix A, with df $= n - 2 = 10 - 2 = 8$, $t_{.025} = 2.306$.

The confidence interval is:

$$67.58 \pm 2.306\sqrt{75.753233}\sqrt{\frac{1}{10} + \frac{(35 - 46)^2}{2474}} \Rightarrow 67.58 \pm 7.74 \Rightarrow (59.84, 75.32)$$

9.85 a. $\hat{\beta}_0 = -13.490347$ - has no meaning since $x = 0$ is not in the observed range. It is the y-intercept.

$\hat{\beta}_1 = -.052829$ - is the estimated change in the mean proportion of impurity passing through helium for each additional degree.

b. For confidence coefficient .95, $\alpha = .05$ and $\alpha/2 = .025$. From Table IV, Appendix A, with df $= n - 2 = 10 - 2 = 8$, $t_{.025} = 2.306$. The confidence interval is:

$$\hat{\beta}_1 \pm t_{\alpha/2}s_{\hat{\beta}_1} \Rightarrow -.0528 \pm 2.306(.00772828) \Rightarrow -.0528 \pm .0178 \Rightarrow (-.0706, -.0350)$$

We are 95% confident that the change in mean proportion of impurity passing through helium for each additional degree is between $-.0706$ and $-.0350$. Since 0 is not in the interval, there is evidence to indicate that temperature contributes information about the proportion of impurity passing through helium.

c. r^2 = R-square = .8538. 85.38% of the total sample variation around the mean proportion of impurity is explained by the linear relationship between proportion of impurity and temperature.

d. From the printout, the 95% prediction interval is (.5987, 1.2653).

e. We cannot be sure that the relationship between the proportion of impurity passing through helium and temperature is the same outside the observed range.

9.87 a. To evaluate the horse breeder's claim, we test:

H_0: $\beta_1 = 0$
H_a: $\beta_1 \neq 0$

The test statistic is $t = \dfrac{\hat{\beta}_1 - 0}{s_{\hat{\beta}_1}} = \dfrac{0.01087 - 0}{.01337} = .81$

The rejection region requires $\alpha/2 = .05/2 = .025$ in each tail of the t-distribution with df $= n - 2 = 7 - 2 = 5$. From Table IV, Appendix A, $t_{.025} = 2.571$. The rejection region is $t < -2.571$ or $t > 2.571$.

Since the observed value of the test statistic does not fall in the rejection region ($t = .81 \not> 2.571$), H_0 is not rejected. There is insufficient evidence to indicate that the gestation period and life span of a horse follow a linear trend at $\alpha = .05$.

b. The form of the confidence interval for β_1 is $\hat{\beta}_1 \pm t_{\alpha/2} s_{\hat{\beta}_1}$

For confidence coefficient .90, $\alpha = .10$ and $\alpha/2 = .05$. From Table IV, Appendix A, with df $= n - 2 = 7 - 2 = 5$, $t_{.05} = 2.015$. The confidence interval is:

$$.011 \pm 2.015(.0134) \Rightarrow .011 \pm .027 \Rightarrow (-.016, .038)$$

Since the interval contains 0, there does not appear to be a linear relationship between the gestation period and the life span of a horse.

9.89 a. $\hat{\beta}_0$(INTERCEP) = −99045
$\hat{\beta}_1$(AREA) = 102.814048

b. To determine if energy consumption is positively linearly related to the shell area, we test:

H_0: $\beta_1 = 0$
H_a: $\beta_1 > 0$

The test statistic is $t = 6.483$ (from printout).

The rejection region requires $\alpha = .10$ in the upper tail of the t-distribution with df $= n - 2 = 22 - 2 = 20$. From Table IV, Appendix A, $t_{.10} = 1.325$. The rejection region is $t > 1.325$.

Since the observed value of the test statistic falls in the rejection region ($t = 6.483 > 1.325$), H_0 is rejected. There is sufficient evidence to indicate that energy consumption is positively linearly related to the shell area at $\alpha = .10$.

c. Since this is a one-tailed test but the output calculates the p-value for a two-tailed test, the observed significance level is:

$$\frac{1}{2}\left(\text{Prob} > |\text{T}|\right) \leq \frac{1}{2}(.0001) = .00005$$

This is the probability of observing our value of t (6.483) or anything larger if $\beta_1 = 0$. Since this probability is so small, there is strong evidence to reject H_0.

d. r^2 = R-SQUARE = .6776

67.76% of the total sample variability in energy consumption around its mean is explained by the linear relationship between energy consumption and shell area.

e. From the printout, for $x_p = 8000$, $\hat{y} = 723{,}467$ (observation 23).

The 95% prediction interval is $(-631{,}806, 2{,}078{,}740)$.

This interval is so large and includes negative BTU's; it is not very useful.

f. We must assume that the new warehouse was subjected to the same climatic conditions as the 22 buildings that were used in the study.

9.91 a. Some preliminary calculations are:

For 6°C:

$$\sum x = 4.8 \qquad \sum x^2 = 9.14 \qquad \sum xy = 124.933$$
$$\sum y = 61.48 \qquad \sum y^2 = 1860.4462$$

$$SS_{xy} = \sum xy - \frac{\sum x \sum y}{n} = 124.933 - \frac{4.8(61.48)}{4} = 51.157$$

$$SS_{xx} = \sum x^2 - \frac{\left(\sum x\right)^2}{n} = 9.14 - \frac{4.8^2}{4} = 3.38$$

$$\hat{\beta}_1 = \frac{SS_{xy}}{SS_{xx}} = \frac{51.157}{3.38} = 15.1352071 \approx 15.135$$

$$\hat{\beta}_0 = \bar{y} - \hat{\beta}_1\bar{x} = \frac{61.48}{4} - (15.1352071)\frac{4.8}{4} = -2.79224852 \approx -2.792$$

The least squares line is $\hat{y} = -2.792 + 15.135x$.

For 12°C:

$$\sum x = 7.7 \qquad \sum x^2 = 23.69 \qquad \sum xy = 352.793$$
$$\sum y = 102.14 \qquad \sum y^2 = 5374.3324$$

$$SS_{xy} = \sum xy - \frac{\sum x \sum y}{n} = 352.793 - \frac{7.7(102.14)}{4} = 156.1735$$

$$SS_{xx} = \sum x^2 - \frac{\left(\sum x\right)^2}{n} = 23.69 - \frac{7.7^2}{4} = 8.8675$$

$$\hat{\beta}_1 = \frac{SS_{xy}}{SS_{xx}} = \frac{156.1735}{8.8675} = 17.61189738 \approx 17.612$$

$$\hat{\beta}_0 = \bar{y} - \hat{\beta}_1\bar{x} = \frac{102.14}{4} - (17.61189738)\frac{7.7}{4} = -8.36790245 \approx -8.368$$

The fitted regression line is $\hat{y} = -8.368 + 17.612x$.

For 18°C:

$$\sum x = 9.5 \qquad \sum x^2 = 34.75 \qquad \sum xy = 496.925$$
$$\sum y = 118.92 \qquad \sum y^2 = 7318.2812$$

$$SS_{xy} = \sum xy - \frac{\sum x \sum y}{n} = 496.925 - \frac{9.5(34.75)}{4} = 214.49$$

$$SS_{xx} = \sum x^2 - \frac{(\sum x)^2}{n} = 34.75 - \frac{9.5^2}{4} = 12.1875$$

$$\hat{\beta}_1 = \frac{SS_{xy}}{SS_{xx}} = \frac{214.49}{12.1875} = 17.59917949 \approx 17.599$$

$$\hat{\beta}_0 = \bar{y} - \hat{\beta}_1 \bar{x} = \frac{118.92}{4} - 17.59917949\frac{9.5}{4} = -12.0680513 \approx -12.068$$

The fitted regression line is $\hat{y} = -12.068 + 17.599x$.

b. The least squares lines appear to be very similar. Since the lines are very similar, it appears the relationship between mean weight gain and rations does not depend on water temperature.

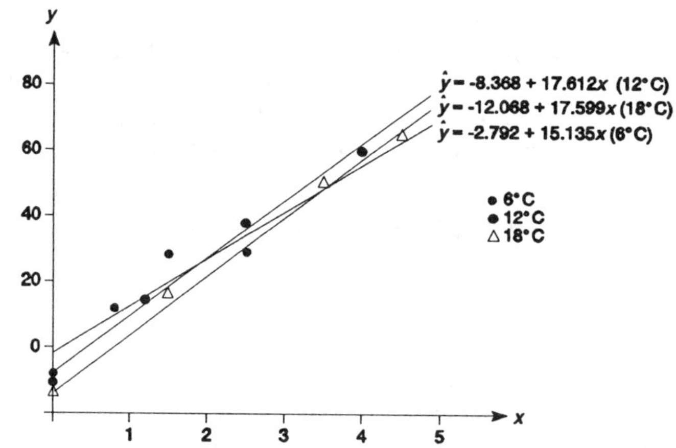

c. For 6° C:

$$SS_{yy} = \sum y^2 - \frac{(\sum y)^2}{n} = 1860.4462 - \frac{61.48^2}{4} = 915.4986$$

$$SSE = SS_{yy} - \hat{\beta}_1 SS_{xy} = 915.4986 - 15.1352071(51.157) = 141.2268104$$

$$s^2 = \frac{SSE}{n-2} = \frac{141.2268104}{4-2} = 70.6134052 \approx 70.6134$$

For 12° C:

$$SS_{yy} = \sum y^2 - \frac{(\sum y)^2}{n} = 5374.3324 - \frac{102.14^2}{4} = 2766.1875$$

$$SSE = SS_{yy} - \hat{\beta}_1 SS_{xy} = 2766.1875 - 17.61189738(156.1735) = 15.67584453$$

$$s^2 = \frac{SSE}{n-2} = \frac{15.67584453}{4-2} = 7.837922265 \approx 7.8379$$

For 18° C:

$$SS_{yy} = \sum y^2 - \frac{(\sum y)^2}{n} = 7318.2812 - \frac{118.92^2}{4} = 3782.7896$$

$$SSE = SS_{yy} - \hat{\beta}_1 SS_{xy} = 3782.7896 - 17.59917949(214.49) = 7.94159119$$

$$s^2 = \frac{SSE}{n - 2} = \frac{7.94159119}{4 - 2} = 3.970795595 \approx 3.9708$$

d. For confidence coefficient .95, $\alpha = .05$ and $\alpha/2 = .05/2 = .025$. From Table IV, Appendix A, with df $= n - 2 = 4 - 2 = 2$, $t_{.025} = 4.303$.

For 6° C, the confidence interval is:

$$\hat{\beta}_1 \pm t_{.025} s_{\hat{\beta}_1} \Rightarrow 15.135 \pm 4.303 \frac{\sqrt{70.6134}}{\sqrt{3.38}} \Rightarrow 15.135 \pm 19.668 \Rightarrow (-4.533, 34.803)$$

For 12° C, the confidence interval is:

$$\hat{\beta}_1 \pm t_{.025} s_{\hat{\beta}_1} \Rightarrow 17.612 \pm 4.303 \frac{\sqrt{7.8379}}{\sqrt{8.8675}} \Rightarrow 17.612 \pm 4.045 \Rightarrow (13.567, 21.657)$$

For 18° C, the confidence interval is:

$$\hat{\beta}_1 \pm t_{.025} s_{\hat{\beta}_1} \Rightarrow 17.599 \pm 4.303 \frac{\sqrt{3.9708}}{\sqrt{12.1875}} \Rightarrow 17.599 \pm 2.456 \Rightarrow (15.143, 20.055)$$

e. We must assume the random error terms (ϵ_i's) are normally distributed with a mean of 0 and a common variance σ^2. Also, the random error terms must be independent.

f. For 6° C:

For $x_p = 4$, $\hat{y} = -2.792 + 15.135(4) = 57.748$.

The 95% confidence interval is:

$$\hat{y} \pm t_{.025} s \sqrt{\frac{1}{n} + \frac{(x_p - \bar{x})^2}{SS_{xx}}} \Rightarrow 57.748 \pm 4.303\sqrt{70.6134}\sqrt{\frac{1}{4} + \frac{(4 - 1.2)^2}{3.38}}$$

$$\Rightarrow 57.748 \pm 54.994 \Rightarrow (2.754, 112.742)$$

For 12° C:

For $x_p = 4$, $\hat{y} = -8.368 + 17.612(4) = 62.08$.

The 95% confidence interval is:

$$\hat{y} \pm t_{.025} s \sqrt{\frac{1}{n} + \frac{(x_p - \bar{x})^2}{SS_{xx}}} \Rightarrow 62.08 \pm 4.303\sqrt{7.8379}\sqrt{\frac{1}{4} + \frac{(4 - 1.925)^2}{8.8675}}$$

$$\Rightarrow 62.08 \pm 10.332 \Rightarrow (51.748, 72.412)$$

For 18° C:

For $x_p = 4$, $\hat{y} = -12.068 + 17.599(4) = 58.328$.

The 95% confidence interval is:

$$\hat{y} \pm t_{.025} s \sqrt{\frac{1}{n} + \frac{(x_p - \bar{x})^2}{SS_{xx}}} \Rightarrow 58.328 \pm 4.303\sqrt{3.9708}\sqrt{\frac{1}{4} + \frac{(4 - 2.375)^2}{12.1875}}$$

$$\Rightarrow 58.328 \pm 5.858 \Rightarrow (52.470, 64.186)$$

9.93 Some preliminary calculations are:

Child	Paranoia	Aggression	d_i	d_i^2
1	7	5	2	4
2	3	1	2	4
3	6	4	2	4
4	1	2	−1	1
5	2	8	−6	36
6	4	7	−3	9
7	10	9	1	1
8	8	3	5	25
9	5	6	−1	1
10	9	10	−1	1
			$\sum d_i^2 =$	86

To determine if there is a relationship between aggression and paranoia, we test:

H_0: $\rho_s = 0$
H_a: $\rho_s \neq 0$

The test statistic is $r_s = 1 - \dfrac{6\sum d_i^2}{n(n^2 - 1)} = 1 - \dfrac{6(86)}{10(10^2 - 1)} = 1 - .521 = .479$

Reject H_0 if $r_s < -r_{s,\alpha/2}$ or $r_s > r_{s,\alpha/2}$ where $\alpha/2 = .025$ and $n = 10$:

Reject H_0 if $r_s < -.648$ or $r_s > .648$ (from Table XII, Appendix A).

Since the observed value of the test statistic does not fall in the rejection region, ($r_s = .478 \ngtr .648$), do not reject H_0. There is insufficient evidence to indicate a relationship between aggression and paranoia in children as judged by this psychologist at $\alpha = .05$.